D. S. Savage was born in 1917 in Harlow, Essex, and educated at Hertford Grammar School and at Latymer School, Edmonton. After working for some years as a clerk in the city he moved to a country cottage to work as a freelance writer and reviewer. He has lived, with his family, in Cambridgeshire, Hertfordshire, Suffolk and Cornwall. In 1947 he received an Atlantic Award for Literature.

Also by D. S. Savage

D. S. Savage

The Cottager's Companion

Illustrated by Gillian Durrant

MAYFLOWER
GRANADA PUBLISHING
London Toronto Sydney New York

Published by Granada Publishing Limited
in Mayflower Books 1980

ISBN 0 583 12850 5

First published by Peter Davies Limited 1975
Copyright © D. S. Savage 1975

Granada Publishing Limited
Frogmore, St Albans, Herts AL2 2NF
and
3 Upper James Street, London W1R 4BP
866 United Nations Plaza, New York, NY 10017, USA
117 York Street, Sydney, NSW 2000, Australia
100 Skyway Avenue, Rexdale, Ontario, M9W 3A6, Canada
PO Box 84165, Greenside, 2034 Johannesburg, South Africa
CML Centre, Queen & Wyndham, Auckland 1, New Zealand

Set printed and bound in Great Britain by
Cox & Wyman Ltd, Reading
Set in Intertype Times

For Mark, Chris and Ted

CONTENTS

INTRODUCTION

When William Cobbett wrote his eminently practical *Cottage Economy* nearly a century and a half ago it was to instruct the farm labourer of that time in ways and means of regaining some of his diminished well-being and independence. The present book has to a large extent drawn its inspiration from Cobbett's manual of self-help. The immense changes of the past 150 years have eliminated the old-style farmer and farm labourer. The countryside has become depopulated while the towns have become crammed with the descendants of the disinherited countryman. There is now coming into being a new type of cottager, the townsman in revolt against an industrial-commercial society, seeking a way of life simple and self-reliant. It is for this 'cottager' that this book has been written.

It has long seemed to me that the real trouble with our present society is that it has lost its human centrality, that impersonal processes are allowed to expand without reference to the real needs of human beings, and no one seems able to stop this happening. The paradox is that nobody really desires the social effects of these processes, although they take them as inevitable, the price to be paid for progress. Progress towards what? So-called progress is running into difficulties, perhaps disaster on a world scale.

A good deal has been written in recent years on the wastefulness of a consumer society making inroads into limited natural resources, and pollution of the environment as these resources are turned into consumer goods. Less attention has been given to the mindless destruction of its own earlier stages by a highly progressive economy. Yet common sense and a calm self-interest should have seen to it that much of the technology of the eighteenth and nineteenth centuries – the development of hydraulics and steam power – was carried forward into the present. This short-sightedness now threatens the further devaluation of man. There may perhaps be ways in which pressure can be brought to bear to check these ominous trends, but such organized effort takes time to make itself felt, and in the meantime the need for men and women to live in as positive and humanly creative a way as they can remains undiminished. A sense of personal accountability for his actions may lead an individual here and there to live in stubborn disregard of the pressure to ant-like conformity, and from such partial detachment from the way of life of the urban collective it is only one more step to move out of the town into the countryside. On grounds of practicality and common sense this has much to recommend it. If collective civilization is heading for a slow, grinding halt it helps neither civilization nor oneself to go under with it.

Cobbett, who took it for granted that country life was superior to the life of towns, and who never spoke of London but as 'the great wen', even in 1821, did not have to argue the point with his readers: they were, of course, countrymen. Nor shall I attempt to do so. If a man is wedded to the town it is useless to try to make a countryman of him. I shall assume that my reader is an actual or intending cottager. I intend to give him, if he thinks he needs it, some useful instruction in the humble and down-to-earth practicalities of day-to-day country living. Compared with Cobbett, however, I am at a double disadvantage in that I can take for granted neither my readers' uniformity of circumstance or needs, nor their basic acquaintance with country ways.

In a time of rapid social change I have felt obliged to write with one eye on present conditions of relative stability and the other on a possible future state of relative scarcity. There will therefore be some unavoidable contradictions and inconsistencies in what I write, since I hope my book may be of use both in the present and the near future. Who knows what tomorrow may bring? – certainly not I, nor, I suspect, the reader. This ignorance applies even to the rather basic matter of the availability of the country cottage. Twenty years ago it was still possible in most parts of the country to rent an empty farm cottage without modern conveniences for a few shillings a week. Today, at least in the greater part of southern England, such places have fallen into the hands mainly of car-owning weekenders from the cities who have modernized them out of recognition and used them as second homes. It is anyone's guess what the position will be in ten or twenty years' time: all we can be sure of is that it will be very different from what it is now. I have taken my aim to be not to instruct the reader how and where to look for a cottage but how to cope with the facts of life when he has got one.

In any cottage way of life the garden is the most important thing. The larger the garden the greater the degree of self-support which will be possible. A cottage which has in addition to a garden, a meadow or orchard, or grazing rights on common land, offers much greater opportunities. An intending cottager should always try for as large a parcel of land as possible when negotiating for a dwelling. Not all cottagers are so fortunate as to live in houses attached to foreshore, lakeside, mountain slopes or forest land, but some do. Even surrounded by arable fields the cottager will have some access to lanes, moors and woodlands with attendant advantages. But however sizeable his holding, he would be unwise to depend upon the sale of its products to provide a cash income. He should rather aim at supplying most of his own needs with a surplus for barter or sale. If he has a spare half or quarter acre on which to grow a specially profitable crop for a cash return – one which might pay rent, rates, mortgage interest or grocer's bill perhaps from the sale of flowers, mushrooms, strawberries, rhubarb or horseradish – he should keep in mind that any year the crop may fail or the market collapse. An independent source of income which links him to the general economy, even if it be a small income, is essential. Cobbett's farm labourer after all had his weekly wage. In today's conditions one is well advised to have a foot in both camps. If the worst comes to the worst the foot firmly planted in the rural world will enable a cottager to feed his family and himself while giving him a base for recovery.

I am not, then, writing for the would-be small farmer, the market-gardener or even the man the Americans call the 'homesteader'. I am writing for the ordinary cottager who may have no more than the average-sized garden but who has the incomparable advantage of living in the countryside rather than the environment of the town. There are many computations of productivity in keeping livestock, cultivating land and the employment of the hours of daylight but few of these will be found to have a useful bearing since individual circumstances such as climate, the nature of the soil and the crops that can be grown are variables. Some notion of the possibilities of large-scale cottaging enterprise may be arrived at if we assume a one-acre garden supplemented by livestock. In terms of additional land the cottager would need roughly:

for 2 goats: $\frac{1}{2}$ ton hay requiring $\frac{1}{4}$ acre land
                          5 cwt grain requiring $\frac{1}{4}$ acre land
for 2 pigs kept 6 months:14 cwt grain requiring $\frac{2}{3}$ acre
                          land
for 12 hens: 8 cwt grain requiring $\frac{3}{8}$ acre land
for rabbits: say, 5 cwt hay requiring $\frac{1}{8}$ acre land
                          $2\frac{1}{2}$ cwt grain requiring $\frac{1}{8}$ acre land

The total, including garden, comes to 3 acres. If we then add another acre for pasture for goats and $\frac{1}{2}$-acre pasture for hens and pigs, all of which can be kept in a more confined area if need be, we find that a holding of about $4\frac{1}{2}$ acres is plenty large enough to produce all the food, except such groceries as sugar, tea, coffee, pepper, salt and spices, as may be needed by a medium-sized family living at a high meat-eating standard and working the land intensively and efficiently. A smaller family eating little or no meat but requiring eggs, milk, butter and cheese would need less land, and still less if they had the means to buy in hay and grain. A semi-vegetarian one would need even less. In any case the vegetable garden must be central to the economy. Here the heaviest labour is digging, the second heaviest harvesting, particularly in man-handling bulky crops like potatoes and apples. There are various ways of lightening the load and in any event digging is mostly done from autumn to early spring when there is little other work. Mechanical cultivators should in my opinion be left to market-gardeners, not only because of capital cost and mounting running expenses but because their use can be justified for only a very short period in the year. The rare cottager with a holding large enough to fall into the market gardening or smallholder class would be well advised to consider animal power – the horse, the pony, even the donkey. In farming there are some indications that the working horse may be making a slow comeback in certain places. The homesteader cottager will find that the working pony costs little to keep. It will live on hay and grass and sleep out of doors all the year round if required although it will be a better worker if stabled and given crushed oats and bran. Donkeys are useful beasts of burden and can be used as draught animals and for light ploughing and harrowing as they are in some 'backward' countries today. Alas, like the goat, they have been looked upon by the pampered and snobbish Englishman as low-caste and therefore ridiculous. Donkeys are, however, very economical and intelligent beasts. They live as long as 30 years and the females can be bred from up to the age of 20, the gestation period being 11 months. They will need half an acre of pasture each with hay and greenstuff in winter supplemented with a handful of concentrate a day. They require no shoeing but must be groomed. And what about the ox? Perhaps the day of the ox may yet return, for a team of oxen ploughs as well as a team of horses and will thrive on a diet of straw, roots and a small amount of meal – the starvation line for a working horse!

Without animals, far more can be done by hard labour than most people suppose, nor is the labour tiring or unpleasant once one's muscles have got used to it and one can go at it at one's own pace. If he has some seasonal help with beet-hoeing, fruit-picking and harvesting, one healthy man can manage a much bigger area of land than the uninitiated would think possible. He can also enjoy a varied diet, a healthy open-air life with no lack of occupation and yet leisure for reading, writing, music and conversation. No well-paid city job could ever give him such a sense of well-being, and in fact the chief danger of this combination of agriculture and horticulture is that it can become too absorbing.

Apart from husbandry there are other matters of concern to the cottager. Some of the subjects I have dealt with in this book may not be liked by some people and I may be asked why I have included a chapter on growing tobacco in view of the hazard of lung cancer, on brewing when the dangers of alcoholism are widely recognized. My answer is that if people will smoke and drink it is better for them to smoke their own tobacco and drink their own beer than that packaged for them by manufacturers. No one need do so if he doesn't want to. It will be apparent that when I describe how to keep and kill a pig, hen or rabbits, I am not writing for strict vegetarians or people who regard all killing of animals as criminal. While I would personally counsel moderation in the eating of flesh foods and would agree with those who point out that it is more economical to feed grain direct to human beings than to human beings through the bodies of animals, I do not regard the domestication and killing of animals as iniquitous. It would not be easy to maintain such an attitude in the conditions of rural life where pests of all sorts have to be destroyed to protect crops and stock, and unwanted male animals slaughtered to maintain a balance and maximum productivity. Many people who would scorn the meat of squirrel or sparrow but would dine on quail, woodcock or hare make this distinction as a result of unconscious social or cultural prejudice. Should our level of culture again regress as it did in the last war we might find ourselves eating not snoek or whale but sparrows, starlings and hedgehogs as was the case here among the rural poor in the last century and as is done in the less developed countries of Eastern Europe today. By general admission the high living standard enjoyed in Britain up to 1975, based on imported foods from primary producing countries, is unlikely to last.

Not only the vegetarian but the sportsman may be offended by some of the material in these pages unless he grasps that I have deliberately taken no account of his point of view. My concern is with the elementary practicalities of life in a cottage. Sport being essentially an activity in which the means employed take precedence over the end pursued, I cannot do otherwise than take a practical view. My instructions for the easy taking of beasts, birds and fishes are given with a view to possible times of scarcity. The cottager, like everyone else, is bound to obey the laws of the society in which he lives; he is bound also to act in accordance with his conscience. In telling the reader how a thing may be done I am not necessarily, even by implication, telling him to go out and do it. I am not saying that under all circumstances, or in particular circumstances, he may do so. This is something he must decide for himself.

No book such as this can tell the intending countryman *all* he needs to know. Limitations of space have compelled me to leave out some subjects which might otherwise have been included, but I have tried to write clearly and concisely about matters which I know from experience are of concern to the beginning cottager.

D. S. S.

Part I

*Friends, Foes
and
Essential Amenities*

CHAPTER ONE

VERMIN

The town dweller pays his rates and in return expects a number of services from Council employees, including the collection of garbage, the disposal of sewage and the suppression of pests and vermin. In the countryside it is different. The cottage dweller with an acre or two of land will have to defend himself against four-footed enemies and winged raiders with an alertness the need for which increases the further he gets from his fellow man. At the edge of a suburb or within a village there may be few enemies. In open country there are predators who take for granted their right to share the contents of his garden and the stores of house and barn. The fox and the rat, as well as the hawk and the crow, may prey on his poultry, chicks and eggs. The rabbit may burrow under his fence and ravage the greenstuff and roots in his garden. Mice and rats will get at his grain and feed stores. Woodpigeons will settle on his kale and strip it to the stump. Blackbirds will eat whatever they fancy from his garden. Sparrows and finches will dig up seeds and peck out the centres from fruit-buds while jays will rip beans and peas from their pods.

First let us separate friends from enemies; and then, among the enemies, those to be outwitted or warned off, and those to be destroyed.

Garden friends to be encouraged are: the hedgehog, the frog, the toad. Among insects: the ladybird, the hoverfly, the ichneumon fly, the lacewing, the centipede (not millipede), the ground-beetle and the devil's coach-horse beetle, despite its name.

Garden foes to be discouraged are: birds of nearly every kind, and besides the harmful creatures already named, moles, voles, badgers, foxes, squirrels and of course, cats and dogs, straying cows, goats and horses. Animals are best kept off by tight gates and fences, and birds by netting where appropriate, and by various forms of bird-scarer, like threads of cotton stretched tightly along the tops of growing seedlings, rattling tins, and foil which flashes and crackles in the breeze. They can also be discouraged by dosing seeds with nasty-tasting chemicals, paints or oil before sowing, and by doing away with conditions which favour the presence of the pest in question.

Insects apart, garden and house enemies to be destroyed where all else fails are: rats, mice, rabbits, foxes, wood-pigeons and whatever non-protected birds pose a threat to food supplies. Non-protected winged predators are: the hooded or carrion crow, the greater black-backed gull, rooks, lesser black-backed gulls and herring gulls, magpies, jackdaws and jays. Animal predators are: foxes, wildcats, domestic cats, rats, stoats and weasels, mink, polecats. The badger, the hedgehog and the mole are classed as predators, yet are all but harmless – and from 1974 the badger is in fact a protected animal.

FENCES. The first thing that a new owner or tenant in the country must do is to go round the boundaries of his property and look at the fences. The best barriers against intruders are brick and stone walls. Oak fencing, open or close, is good, the protection it gives depending on its height and thickness. The most pleasing boundary is a live hedge. Quickthorn and holly are the most stockproof, but beech, lime, hazel, maple and hornbeam all make stout hedges of quick growth. Plant a row of healthy seedlings 8 in. apart in the dead of winter, manure them freely the first year, then cut them right back in December. Trim the sides of the new hedge as it grows, leaving the top shoots to grow to the required height before heading. A wire fence may be needed while they are growing. Don't use barbed wire. A plain fence of wire strands tightly strained will keep out sheep and cattle. The end or corner posts must be staunch; make them rigid with diagonal spurs or struts to take the strain and see that they are well pickled with creosote or tar. They may then be embedded in concrete or in a socket made by four large stones. Wood fence posts should always have their bases charred in the fire and coated with tar or pitch to prevent rot. A cheap fence is split chestnut paling, sold in rolls and erected by wiring it at intervals to iron stakes sledged into the ground. There are several kinds of chain-link and welded-mesh fences, some with small mesh at the bottom to keep out rodents and wide mesh at the top. Generally the price of such fences is in inverse proportion to the size of their mesh: i.e. the smaller the mesh the more they cost per yard.

Where rabbits abound, firm measures are needed to keep them outside. Unless defended by brick or stone, your garden needs to be enclosed by wire netting of 1-in. mesh, at least 3 feet above ground level, with 6 in. or more buried to prevent the beasts from tunnelling under it. Guard against baby rabbits as well as adults. A good wooden fence need have the netting only at its foot – use 2-ft high netting, 1 ft being let into the ground and the rest stapled to the wood on the outside of the fence. The same can be done with fences of other types. With living hedges the netting is fixed independently on the inner or garden side, the top edge being bent over outwards and so fixed with brackets to hinder rabbits from climbing it. Rabbits hate onions, so a border of any of the onion family all round the garden will help keep them off it. Like birds, rats, mice, moles and voles cannot be kept out by any sort of hedge or wall. No fence is stronger than its weakest point. When making a wire-netting fence, as in chicken-runs, get some plain No. 8 gauge wire and interweave one strand at the top and one near the bottom throughout its length. These may be stretched taut without damage to the netting, which itself cannot be stretched or tightened. A third wire will strengthen the netting against stock. A gate in the fence must be kept shut always, and a solid concrete or timber sill laid on the ground exactly parallel with its bottom and so fitted that there is no room for a rabbit to squeeze through. Stoats and weasels will attack young chickens if they can get at them through ordinary 2-in. wire netting. Chickens killed at night may be the victims of a rat which has taken up residence inside the hen-house; it must be found and killed without delay.

FIREARMS. A gun may be the only way of keeping down rabbits, rats, woodpigeons, stoats and weasels which have become serious pests. Guns used are the air-rifle, the ·22 rifle and the double-barrel 12-bore shotgun with 26-in. or 30-in. barrels. No licence is needed for an air-rifle on one's own property, but a Firearms Certificate must be obtained from the local police for a rifle or shotgun even when its use is restricted to the owner's own garden and curtilage. The air-gun is not always effective and the bullet from the ·22 rifle is too dangerous inside a mile range. This leaves the shotgun, and if used only for shooting in the garden and yard then the smaller make of ·410 calibre, the 'four-ten', is adequate. Some specially light 'garden guns' fire cartridges the tubes of which are made of a special paper which separates from the metal head on firing and travels through the bore of the gun intact to disintegrate from the muzzle, leaving the shot to travel on. These are especially suited for use in confined spaces where a more powerful gun would cause structural damage. Killing pests is not a form of sport, so sitting targets should be chosen; to shoot on the wing may mean wounding a bird which must then die a lingering death outside the garden.

Rabbits in the garden are elusive, dodging about under cover in the vegetable patch. Try for a sitting shot in early morning and just before dusk.

Rats are easier to hit. They will climb up bean- and pea-rows at any time of day, but are best hunted in the evening. In winter they move into buildings, migrating in summer to burrows in hedgebanks. Their burrows have smaller entrance holes than rabbits', and their dung, usually found near these holes, resembles the rabbit's but is elongated and dark in colour. They can be attacked in their holes by ferrets, who drive them out for dogs to kill.

SNARING RABBITS AND RATS. The use of snares is described on page 168. It is not so easy to see the rabbit's tracks in garden soil as on turf, but if the rabbit is 'guided' along a low, wire-netting fence with a wire set in each opening, he will be caught. Snares may be set for rats, but where rabbits require a three-strand wire and a 4-in. noose, the rat snare need be of only two strands or one, with a 2-in. noose. Refer to the passage on snaring rabbits; set up the same sort of snare for rats but add to it a 'bender' in the form of a stout hazel rod about 4 ft long, to the top of which is tied the string of the snare. Fasten a piece of wood with a transverse notch to the string close to the edge of the wire. Drive a peg with an engaging notch slantwise into the ground. Bend down the hazel switch and engage the notches so that they will come apart at a light pull on the wire. Then set the snare, with a little cleft stick, across a rat-run or at the mouth of a hole. The rat runs his head into the noose and pulls apart the notched pieces, the bender flies upwards and jerks the loop tightly round his neck and the wretch is neatly gallowed. A snare of this pattern, able to lift 18 lb weight, will do for foxes. Foxes are wary foes, and if one has to be dealt with it is best to pursue several methods – shooting, snaring and trapping – at once.

RATS, MICE AND MOLES. Rats and mice may be caught in patent traps of the spring or cage type. And they may be poisoned. It is illegal to poison rabbits. Rats may be gassed by connecting a hosepipe with the exhaust of motor-car or mower and running its other end to one of the holes in a burrow, the other holes being first stopped with earth. Start the engine and let it run for a while and the rats will be killed by the lethal gases which will filter through the whole run system – clear proof of the noxiousness of the fumes which we put up with in our city streets. Moles can be driven away from their runs by this means, if their exit holes are left unstopped. Smoke cartridges against moles may be bought from country-town ironmongers. A large onion cut in half and wedged in a mole run is said to send them away, since moles cannot abide strong stinks (as they are blind, it cannot make them weep). The village mole-catcher, who trapped moles and made up their downy skins into men's weskits, is now a figure of the past but mole traps may be bought from the ironmonger in most market towns.

Once rats have got into a building they are difficult to dislodge. Make sure that all walls of a house are rat-proof, and look at all air-bricks and metal gratings which ventilate its foundations, as well as openings under eave-boards, and put them to rights. Rats will get into thatched roofs, hence the wire netting sometimes used to cover such roofs to keep them out. Wooden sheds and outhouses must be protected at ground level by brick or concrete footings. If these are lacking, the remedy is to fix small-mesh wire netting all round the walls, sunk one foot into the ground and rising one foot above it, stapling the upper part to the side of the shed. Rats may be prevented from gnawing a gap at the bottom of a door by a strip of zinc or sheet-iron screwed along the outer side, coming down to within half an inch of the sill. The danger from rats in case of social breakdown and plague should be considered, and steps taken now to make all one's buildings rat-proof.

Where rats are attracted to kitchen waste in compost boxes they can be poisoned by a sprinkling of gypsum on the rubbish, or deterred by chloride of lime. Mice can be kept from food by the crushed leaves of ground elder.

Inside or outside, rats and mice are best dealt with by a not too well-fed cat, preferably female, of the old-fashioned hunting strain. A fox terrier will deal effectively with the biggest rat. These creatures are more efficient than traps, baits and poisons; and every countryman knows that no home is complete without a cat and a dog.

INSECT PESTS. A hedgehog brought into the house will kill and devour all cockroaches. Or they can be got rid of by strewing on the floor before bedtime a mixture of crushed gypsum with twice the quantity of sugared oatmeal. Beetles can be poisoned with a mixture of equal weights of red lead, sugar and flour. Bluebottles: keep them away from the meat safe by placing a saucer of water and permanganate of potash near its door. Mosquitoes: drive them from the room by holding a lump of camphor gum over a lamp chimney.

Flies: attract and destroy them by strong, sweet green tea set about the room in saucers. Ants are repelled by the herbs pennyroyal, spearmint and tansy. Destroy ants by dropping quicklime on their nest entrance and washing it in with boiling water. Carpet- and clothes-moths are repelled by the scent of dried lavender. Clothes in a press may be sprinkled with dried leaves of lavender, mint, rosemary, sage and wormwood to keep the moth away.

CHAPTER TWO

STONE WALL AND QUICKTHORN HEDGE

It sometimes happens that part of a country property has a stone wall in need of extension or repair, or a thorn hedge that has been let go and must be laid, or layered, to make it animal-proof.

STONE WALLS (Fig. 1) are found in places where a rocky substrate lies close to the soil surface, with outcrops and loose stones here and there. Walls around fields have usually been built originally with stones lifted from the field itself, supplemented by stones from a nearby quarry. When building a dry stone wall – 'dry' because no mortar is used – the first thing is to sort out the stones into sizes and to arrange that supplies shall be carried or drawn by sledge downhill rather than up. Treat walling as a summer rather than a winter job.

Peg a double line where the wall is to run and dig a 6- or 8-in. deep foundation trench, paving it with stones a bit wider than the base of the wall at ground level. Tools needed besides spade and line will be an expanding rule, a 4-lb club hammer, a cold chisel, a mallet, an iron crowbar – with a thick 6-ft long plank and a length of rope if heavy stones are to be hauled and levered into place.

Walls are built with two parallel rows of stones fitted closely together, and the rule is to lay each stone on its broadest face, lengthwise into the wall, placing the larger stones in the lower courses, the size of the stones getting smaller towards the top of the wall, the wall being 'battered' (sloped) each side at an angle of 10 degrees. It follows that a wall with a 34-in. base will be 20 in. wide at 3 ft 9 in. high, or 14 in. wide at 6 ft. Such a wall can be built up at any time without having to be pulled down and rebuilt. If the wall is a long one, fix into the ground, at each end, timber frames the exact shape of the wall to be built, with guidestrings stretched from one to the other, and work within the strings,

Not more than two foundation stones make the width of the trench. Follow them with further layers, each sloped inwards a little, and so placed as to cover the join in the lower course. Work from each side in turn and fill in any space between the two sides with small rubble packed tight. The art is to select stones that will fit into each other, with a smooth face on the flank of the wall. At about 21 in. from the ground, large rectangular 'tie-stones', as long as the wall's width, must be placed across it at 3-ft intervals, to hold the two sides together. The course is made up with smaller irregular stones and the building continued as before. If the wall is to be 6 ft high, then further tie-stones will be put in at about 4 ft. Whole stones are best, but if a tie-stone has to be *cut*, cut it by scoring a groove along all

Fig. 1

four faces with a cold chisel and hammer; then, placing a board upright over the groove, strike it a few blows on its edge with the mallet: stone fresh from the quarry is the easiest to work. A 4-ft wall can be finished by covering it at about 3 ft 3 in. high with flat single stones and then *capping* it with flat vertical stones of roughly equal size stacked tightly edgewise. Put in a heavy stone here and there to hold them in case some are knocked out. Go along the finished wall and hammer small stones into all gaps. If skilfully built, such a wall will stand for decades without attention. Where stone abounds, the art of walling can be applied to the erection of large or small buildings, stones being raised to the upper levels by block and tackle or the use of the high-lift on a tractor.

Where stone is scarce, concrete blocks, tongued and grooved for rigid fitting together, may be made in a wooden mould and used instead. As an alternative gunny-sack halves may be stitched into narrow bags and filled with a sand/cement mixture, then stacked up dry to make the wall. A heavy rainstorm will set them hard, or they can be hosed or doused with water as the wall is built.

LAYING A HEDGE (Fig. 2). A field hedge often has a drainage ditch on the outer boundary; and the winter work of laying a thorn hedge so that it grows in such a way as to be proof against stock is done where possible from the ditch side. The hedger wears mitts of tough leather and uses a billhook to sever the wood and a mallet to knock in any non-live stakes that may be wanted. The work is essentially that of selecting, bending and weaving the layers semi-horizontally and in one direction between the vertical, dead or living, stakes, the job being finished off by binding or braiding the top of the hedge by headers – long lengths of live or dead stem about $\frac{1}{2}$ to $\frac{3}{4}$ in. thick.

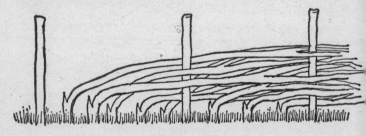

Fig. 2

Without doubt the best hedge is an all-living one in which stakes, layers and headers are all selected and trimmed from the live wood. But it is quite usual to have only the layers live, cutting the stakes and headers from waste hedge material or buying them from the owner of a coppice in bundles holding enough to do a chain at a time.

Proceed like this: 1. First clear the hedge bottom of brambles and other rubbish, cut out unwanted growth at the sides and chop out all old, useless stumps from the hedge-row. 2. Thin the hedge, not too severely at first, leaving un-touched all stems required for stakes, layers and headers. 3. For stakes, select the thickest stems at about 3-ft intervals and top them at hedge-height. 4. Then work along the hedge, with an upward blow of the billhook cutting each layer as close to the ground as possible and bending it away from the cut, which should be clean and long *but only deep enough to bring the layer down*, otherwise sap-flow may be impeded and the layer die. The aim should be to get the hedge to lean slightly over away from the ditch, with the 'brush' or small wood pushed through on to the field side where it will pro-tect from grazing cattle the young shoots that will later break from the bottom of the hedge. Lay the thorns with an upward slant, bending the lateral growth on them over the other way and weaving it into the fabric of the hedge. 5. If live headers are to be used, leave some very long thin stems every 10 or 12 ft on the ditch side, and when a section of laying is done, cut them upwards at the bottom, like the layers, but bend them in the contrary direction, and braid them between the stakes along the hedge top, where they will both strengthen the hedge and hold the layers in place. If dead headers must be used, cut or buy hazel rods 12ft by $1\frac{1}{2}$ in. at the butt end, or use strong brambles. Key in the end of each rod so that it cannot work free, and weave it in and out as part of a continuous plait. The best hedge is one which is thick at the bottom, tapering towards the top; and cutting the layers low will aid the growth of shoots from the hedge bottom. Fill a large gap by pegging a layer down to the earth so that it may take root or grow into new stems. At the same time take into account the possibility of keeping here and there any good oak, elm or ash saplings which might grow up into tall hedgerow trees. On finishing the hedge, clean the ditch and throw the soil along the hedge bottom.

Trimming the hedge to keep it in shape is done at odd

times throughout the year. Using a slashing hook, trim the hedge with upward sweeps on the sides and with downward strokes at the top.

ELECTRIC FENCES can be a great help when animals are to be contained in a field. They are light and easy to move, and as almost no current is lost from the live wire except when it is earthed by contact with an animal, the same small battery can power an unlimited length of fence. A 6-volt motorcycle battery is used and will last for several months, or the current may be got from the mains. Stranded 24-gauge or 14-gauge plain galvanized wire makes the fence, but any sort of plain wire can be used. Horses and cattle need only one wire to control them; pigs need two, and goats three – but the third, bottom wire need not be live.

The tackle needed to fence a $\frac{1}{2}$-acre field for goats is: 1 fencer unit, with battery, 1 mounting post, 4 corner tripods, 8 $\frac{3}{8}$-in. fencer posts, 16 adjustable insulators, 6 corner insulators, 2 wire strainers, 1 plastic gate handle. For a 1-acre enclosure an extra 8 posts and 16 insulators are needed.

The corner tripods are first set out, the two supporting props standing inwards to take the strain of the wire. The metal mounting post is knocked into the ground and the switched-off fencer unit fixed to it and the battery connected. (The metal post 'earths' the current: if mounted on a wood post a connection to an earth terminal must be made.) The fence posts are then set out at 12-yard intervals and the top and middle wires attached to them by the adjustable insulators, the circuit being completed back to the unit. The top wire should be about 30 in. above the ground, at the goats' eye-level; the second wire some 18 or 20 in. and the third, dead wire, 9 in. above ground.

A gate in the fence (Fig. 3) is made by fixing, about 2 ft apart, two rigid posts which will take the strain of the fence, by cutting the wire, slipping on the plastic gate handle, then looping the end of the wire so that it can be hooked over to re-connect the circuit at the other side of the gate. (The 'gate', that is to say, is just two strands of detachable wire. The gate itself does not take the strain of the fence – if it did,

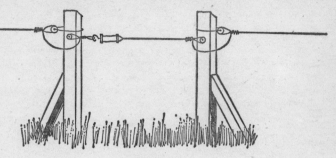

Fig. 3

the fence would sag every time it was opened.) The fence wire is now pulled taut by turning the strainers, a tight fence being necessary for the making of positive skin contact through the animals' hair should they touch the wire. The gate is unhooked, the goats led through, the gate fastened and the current switched on. To train goats to respect the fence, bait the wire with green fodder, so that when they first take it they get a shock through the mouth. As the current is not continuous but intermittent, at some 40 impulses a minute, it cannot harm them. An electric fence is often used to fold animals on land for controlled grazing of kale, etc., the wire being moved daily to take in a fresh strip of grazing. Rabbits can be very effectively kept out of a field by an electrified fence set 4 in. from the ground.

HURDLES. The cottager who keeps animals of any kind will find it useful to have a dozen sheep hurdles on hand for use when required. They are cheap and take up little space when stacked. A pen for animals – say, for a kidding goat – is quickly made out of four or eight hurdles and four corner posts. Sheep hurdles have a loose iron loop at the top of one end for slipping over a post or stake, and pointed footings which grip the ground; they are surprisingly firm when put together. They make useful temporary gates or fences, and can be rigged up into shelters with roofs of corrugated iron or of bracken or heather thatch.

CHAPTER THREE

THE COUNTRYMAN'S TOOLS

Garden and woodwork tools apart, the countryman will find that he needs some of the following implements from time to time.

Axe: English type, with broad head and ash shaft. For felling trees, splitting logs. In dry weather put the head in water for 20 minutes to swell the wedge and keep all tight. Hone edge with whetstone, sharpen with file.

Bitel: Two-handed wood mallet for driving in hedgerow stakes and knocking wedges into timber. The head is of applewood held by a forged iron band at each end, the handle of straight-grained ash tapered so that it fits tightly into the head.

Beet knife: Slightly curved narrow blade with a downward spike at one end and wood handle at the other. The beet is picked up by the spike and cut with the blade.

Besom: A yard-broom made by tightly binding a bunch of long twigs in two places near the butt end, trimming the butt with a hatchet, and driving into it a pointed peeled stake as a handle.

Billhook: Broad-bladed heavy steel chopper for hedging: will also serve for cleaving firewood.

Club hammer: For driving metal stakes into the ground, or for masonry work.

Crowbar: 4-ft steel bar, chisel-pointed at one end, crooked and notched at the other, for pulling out nails, opening cases, making holes for stakes, levering heavy stones.

Hay rake: Large wood rake with wide-spaced teeth for collecting hedge-trimmings, leaves, grass.

Pickaxe: One end pointed, the other chisel-edged, for excavating in subsoil, clay, rock.

Rammer: Home-made from part of a tree. 3 ft 6 in. long, 5 in. across at the base and tapering to form a hand-hold. For firming earth round posts and poles.

Scythe: For cutting grass, harvesting corn.

Sickle: See p. 86.

Swap-hook: Sickle-like blade slightly offset from handle, for cutting grass, nettles, weeds. (A sandstone or carborundum rubber is needed for keeping the edge on swap-hook and scythe.)

USING A SCYTHE. Though hardly an implement for a woman, the scythe (Fig. 4a) is less difficult to wield than its formidable blade suggests. Scythe blades come in sizes from 24 in. to 40 in., the average being about 34 in., and the wood shaft or 'snaith' varies in length, to suit the height of the user. The snaith of the traditional English scythe has a double curve which adapts well to the hand positions and body movements of the user. With practice it becomes a most efficient machine for cutting grass, coarse growth or

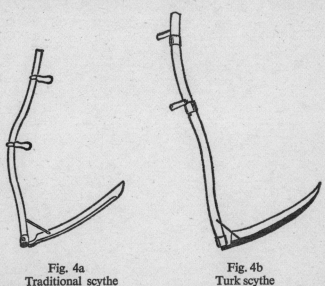

Fig. 4a
Traditional scythe

Fig. 4b
Turk scythe

corn: but tufted meadow-grass is resistant to it. Besides the English scythe there is an admirable light scythe of Austrian manufacture called the 'Turk' (Fig. 4b).

When cutting with the scythe, stand as straight as possible, your legs about 18 in. apart, left foot a little behind right. Grasp the handles or 'doles' of the snaith and begin the stroke with the right hand just before the right leg, then bring your right hand swiftly around to your left knee. The blade will seem to swing itself forward and then swing back for the next stroke. The art of scything is to get into and keep up a regular, rhythmic, unhurried motion, without jerkiness, not letting the blade point dig into the ground. Do not attempt too much at first. With practice your scything will improve until it becomes second nature. To sharpen the blade, wipe it over with a fistful of grass, put the butt of the snaith firmly in the ground and start whetting at the base of the blade, working slowly towards the tip and holding the blade firmly by placing the left hand on top of it. To test the blade for position should the wedges loosen: hold the snaith in the right working position and see that the blade is lying flat along the ground. Tighten the wedges. Carry it over the shoulder, blade at the rear.

LIFTING SACKS. To tie a sack, pull the mouth lengthwise, stretch it and fold into pleats. Twist the twine twice round the pleated mouth and pull tight, then twist round once and pull tight again. Tie with one extra twiddle, and knot. In lifting a sack, first get it on to a raised platform, then back up to it and take the weight as high up on the shoulders as you can.

A heavy sack or other weight can be hoisted by means of sheers (Fig. 5) – two poles lashed together near the top, then stretched apart at the bottom. The weighty object is tied to the top of the lowered sheers, which are then raised by pulling on a line. A third pole can be positioned to make a tripod.

Sheers making use of four lines can both raise weights and move them for a distance in a series of three operations (Fig. 6).

Fig. 5

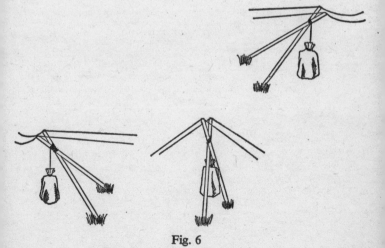

Fig. 6

When heavy objects need to be hoisted, a block and tackle may be used where a suitable beam can be found to take it. A locking rope tackle can be made to lift a ton limit. Using slings the locking rope tackle may be employed horizontally if suitable stout posts or trees are available, the slings being looped about the tree trunk and the top hook of the tackle fixed in them so that heavy objects may then be dragged across the ground, if necessary assisted by rollers and drawn by the moving block of the tackle.

A very large boulder may be moved by a long lever and the use of graded pebbles, small ones being inserted when the leverage is applied, the fulcrum being adjusted and the advantage being exploited by the insertion of larger pebbles until the boulder is rolled over, the operation being repeated. The device called the Spanish windlass, or tourniquet, most familiar from its use to tighten the wooden frame of the fret-saw or box-saw, and which consists of a rope with a rolling hitch and a handspike in the bight for leverage (Fig. 7), can be used for loosening a stubborn boulder or stump.

For *garden tools* see p. 66 and for *woodman's tools*, p. 52.

Fig. 7

CHAPTER FOUR

WATER SUPPLY

A country cottage without an adequate supply of drinking water would be no use to anybody, but there are few such cottages in Britain because of the heavy rainfall – from well over 80 in. in some parts of Scotland to below 30 in. in East Anglia. Rain water can be caught, stored and used in the rare cases where there is no usable water below ground, so let us deal with that first.

RAIN WATER. As one inch of rain will yield $\frac{1}{2}$ gallon from each square foot of roof, with a roof area of 1,000 sq. ft and a rainfall of 35 in. a year, 17,500 gallons will be collected. Some of this will be wasted, but about 85 per cent should be usable if properly stored. If a net 15,000 gallons are available for use, this works out at roughly 40 gallons a day throughout the year. Most town households (1974) use about 33 gallons of water daily for all purposes, including bath, laundry, garden and car; so 40 gallons would be more than enough for a small country family.

Where the house is small, the roofs of outhouses and sheds must be brought in to feed the tank, or a ground-level rain trap constructed, sloping slightly towards the tank. If the water is for drinking, all lead on the roof should be taken off and replaced by copper flashings and plastic rainwear. Slate roofs, being non-absorbent, are better than tiles.

A storage tank will be excavated to your instructions by a contractor near one end of the house, or it can be dug by hand, using pick, shovel, bucket and wheelbarrow. Directions for making an underground cistern of 18,000-gallon capacity, as well as many other practical building details, are given in *A Manual on Building Construction* issued by Intermediate Technology Development Group Ltd, 25

Wilton Road, London SW1V 1JS. As water is heavy, the floor of the tank must be well supported and the sides made quite watertight both to prevent leakage and to keep out impurities. The tank should be covered in, and have a manhole fitted large enough to allow it to be cleaned: a manhole with a turned-up edge, having a cover with a turned-down lip to fit over it, will keep out surface water. A circular tank is cheaper to build than a square one.

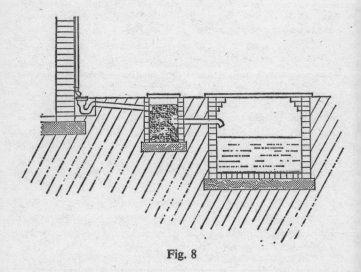

Fig. 8

If used for drinking, rain water should be run through a filter chamber placed between supply-pipe and tank (Fig. 8). Fill the chamber with a 2- or 3-ft square bed of washed gravel 1 ft thick, on top of which is a $1\frac{1}{2}$-ft layer of fine sand. Replace the contents every year.

WELLS. An old cottage may have a well already sunk into the water-bearing stratum of the soil. It may be forgotten and invisible, covered with a stone slab, earth and grass, or its presence may be betrayed by a rusty iron pump. Uncover it and inspect the state of its brickwork or 'steining', and if you are doubtful, send some of the water to a public analyst for testing, meantime boiling all water before drinking it. Wells are either shallow or deep, drawing water either from the subsoil or from a water-bearing stratum lying beneath an impermeable one, often at great depth. Shallow wells are more likely to be polluted, especially if they are near drains, cesspools or cattlesheds. If a well is found to be polluted, and the source traced and removed, it can be disinfected by adding a 5 per cent chlorine solution in one gallon of water to each 1,000 gallons of water in the well, leaving it to stand for two hours, then pumping the tainted liquid to waste.

FINDING WATER. If there is no well, or the old well is unusable, a new well may be dug by a local well-digger with the right equipment, or by two or three unskilled men in a few days. The driest part of the summer should be chosen, when the water level is low, giving a realistic idea of how much is available, and when the soil is easier to shift. The presence of water can be inferred from a knowledge of local geology, but the time-honoured method of locating a water supply is through the practice of divining by means of a dowsing wand or twig.

No special skill is needed for dowsing, though sensitivity varies from person to person. A forked piece of springy wood is used, freshly cut from a live tree, with arms or handles about 18 in. long and $\frac{1}{2}$ in. thick. Hold the end of each handle with palms upwards, knuckles slightly inward, the arms unbent and a little below the level of the hips, and walk slowly about the area of search until the far end of the stick dips suddenly downward, indicating water. A practised dowser will follow an underground stream or a number of converging streams to a blind spring known as a 'head' or 'knot', where his rod will become agitated, and there give directions for the digging of the well. These blind springs seem to have an attraction for animals, and dowsers claim that gnats dance always over blind springs or nodes, that moles make their fortresses on them, and that unconfined geese and hens always find blind springs on which to make their nests. But bear in mind that no dowser can tell with certainty how deep underground is the spring he has traced, nor can he vouch for its suitability for drinking.

TYPES OF WELL. An artesian well is one where no pumping is needed to bring up the water, the upward force being imparted by the weight of collected underground water just above the point where the well shaft enters the natural reservoir it taps. Such a well can be sunk on your land only if there should happen to be a basin formed of a layer of porous rock lying between two impermeable layers – usually clay – the porous layer curving upwards to places at each end where it takes in moisture which then sinks through and collects at the centre of the basin. Such wells are drilled, not dug. The specialist will have mechanical tools which can penetrate to any depth. For depths of up to 35 ft, well-boring kits can be obtained which can be worked manually. They consist of an auger made to take a number of separate cutters for use in different soils with a bailer for scooping loose sand from the bottom of the hole, further driving tubes and connectors can be added as depth increases. After water is struck, the bore hole is sleeved with continuous pipes knocked in on top of one another. In true artesian wells the force of the water is often enough to carry it to the storage tank at the top of a building without pumping.

Underground springs can be tapped in much the same way by bores of 4–6 in. in diameter, although here the water will have to be pumped to the surface. For a very small demand, the Abyssinian tube well may meet your need. This is made up of strong tubes of mild steel which are driven into the soil, the bottom length perforated to let in water, having a hardened steel point for penetration of gravels and clays. As each length is knocked in, a further length is screwed or collared on to it.

DIGGING A WELL. To dig a well manually, pick and shovel are needed, with a tripod of stout poles rigged up with rope, pulley and hook and a bucket for hauling up the excavated earth below shoulder level. Practised well-diggers excavate depths of about 3 ft at a time, timbering each section with pieces of board to prevent earth falling in. They dig until they get to several feet below the lowest level of the subsoil water, keeping the well dry meantime with continuous pumping. At the bottom they make a sump for the temporary pump by sinking a 6-in. diameter pipe into a hole, laying around this a concrete floor. They then build up the walls from the bottom, dismantling the timbering as the walls rise. Old wells are generally brick-built, with gaps at intervals several feet up from the bottom to allow water to gush or seep in. If sections of concrete sleeving are used instead, the lower ones must be so arranged as to permit a similar inflow of water. Higher up, however, the brick wall will be built solid with cement mortar backed with clay to keep out surface liquids. At the top the diameter is reduced by corbelling to about 2 ft, rising not less than 1 ft above ground level. The ground is then thickly paved or concreted, the paving sloping away from the well to a distance of at least 4 ft in every direction. When the well is finished the temporary pump is removed and the sump in the floor filled in, a permanent pump being now fixed in place with its suction pipe running down to the water level.

The retaining wall of the well may be of stone or of wood. Elm will last for centuries in waterlogged condition without rotting. If wood were plentiful and bricks scarce the well shaft could be timbered – boxed-in sections placed on top of one another, the space between the wood casing and the hole being filled with gravel or graded stone rubble. Oak casks with top and bottom knocked out would seem to make a good lining, as would large steel drums of uniform size thickly coated inside and out with bituminous paint.

PUMPS. The primitive bucket and rope in wells was replaced by the iron pump in the last century. The mechanical pump (Fig. 9), made of wood or iron, served man for many generations. A suction pump operates by the creation of a temporary vacuum and the emission of water through an outlet valve. Such pumps work well in shafts up to 25 ft deep, trouble arising only when the valve seating or piston becomes worn. Worn pumps need to be primed with a jug

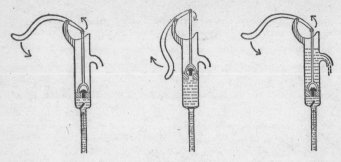

Fig. 9

of water before use. This type of pump is still made. Pumps which will raise deep water are more complicated since the lift can no longer be got by suction from a piston at ground level; the piston must be placed at water level and connected to the handle of the pump by reciprocating rods. The *Royale* deep-well pump, now made by Pompes Briau, of Tours, France, is of this type and will operate from depths up to 130 ft. H. J. Godwin Ltd, of Quenington, Cirencester, Gloucestershire, also make hand pumps, one of which, wheel-operated, will lift water from 320 ft.

SPRINGS. A surface spring of pure water near the house makes a well unnecessary. Deep springs are best, their water having been thoroughly filtered in its descent through soil, sand and gravel. They are less likely to run dry in summer than surface springs. Water from such a spring on a hillside above the house can be piped into an outdoor storage tank big enough to contain at least a week's supply of 200 gallons. While no doubt the best tanks are of slate slabs slotted together, a useful one can be built with concrete blocks rendered inside with cement mortar and outside with impervious clay puddle. It must have an overflow pipe, a drainage tap and a cover. Just as satisfactory, perhaps, and quicker to erect would be a modern swimming pool with rustless frame and plastic liner. From any such tank water can be brought to the house through metal, rubber or plastic tubing by means of a length of copper pipe bent into a siphon and fixed with its end some inches from the bottom of the tank; screen the opening to keep out unwanted dirt and insects. If the spring is too low to feed the tank by gravity, the water can either be pumped up to it or pumped directly to a tank in the roof of the house.

While this could be done by an electric or wind-driven pump, the most effective engine for it is the hydraulic ram, by means of which the energy of a quantity of water with a small head is used to lift a proportionate quantity to a higher level. The way it works is that water flows from the source – a spring or a stream – along the drive pipe and through the waste valve of the ram. A waste outlet is therefore necessary. The rapid flow of water through the open waste valve causes this to shut. The momentum of water in the drive pipe then forces some of the flow through the delivery valve. This compresses the air cushion in the domed air-vessel and causes a continuous outflow through the delivery pipe to the storage reservoir. If there is a sufficient fall of water, rams will work from springs giving as little as one gallon a minute; with a heavy rush of water, a fall of less than 3 ft can be used, lifting water to heights of up to 400 ft. Rams are trouble free, cost nothing to run, and last indefinitely: and they are relatively cheap.

MODERN PLUMBING is so labour-saving that it would be short-sighted not to install it in a cottage where it is lacking. The amateur can manipulate some modern materials without plumbing skill. Heavy-duty polythene or alkathene tubes, which can be cut with knife or saw, have thick enough walls to allow connection by special screwed or compression fittings. Begin by fitting a tank under the roof, with a ball-valve and waste-pipe and an outlet to supply the pipes running to bathroom and kitchen. Polythene is elastic and will expand without bursting in a freeze and thaw, and is especially suitable for cold water supply. But as it has a melting point of only 225–235°F it is not suitable for hot water.

CHAPTER FIVE

GARBAGE DISPOSAL, DRAINAGE, SANITATION

In a properly managed country household the disposal of garbage presents no problems, most of the waste being returned to the soil either indirectly through animals or directly through composting. The housewife should have four lidded bins for the main types of refuse: (1) food scraps, (2) vegetable waste, (3) metal, glass and plastic trash and (4) paper. Of these, (1) and (2) will be emptied daily, (3) and (4) when full. (1) can be fed in a mash to pigs or poultry. Otherwise the scraps can be composted. To avoid attracting rats, cover with earth and sprinkle with chloride of lime, or use a patent 'Rotol' lidded plastic compost bin which requires no activator and has a quick turnover of compost. (2) can go straight on to the compost heap, (3) must be buried in a pit which can be earthed over when full, all metal cans being first crushed flat, (4) can be burned on the kitchen stove or in a garden incinerator. This may be of the heavy brazier type on legs – avoid flimsy wire baskets which burn out and corrode – or it can be built up with loose bricks and a metal grid having a space beneath for ash removal. This will burn not only uncompostable paper but any organic material too dense to be rotted down.

DRAINS. Until recently old cottages and farmhouses had no plumbing and drainage, the kitchen water in bucket or bowl being flung out of the back door on to the garden, and baths carried out and tipped into a ditch. Rather than let waste water escape from the house to run on the soil surface, dig a trench not less than 1 ft deep and with a fall of about $\frac{1}{8}$ in. in the foot through towards the end of the garden, line it with medium-sized stones, gravel and ash and place on top a line of agricultural drain-pipes loosely butted together, with a sump or soakaway at the far end, and near to the house a grid to strain off large particles which might clog the pipes. Such a drain might be laid in part beneath a garden path, and run towards a flanking ditch.

DISPOSAL OF EXCREMENT. By and large, city-bred people have come to consider their own body wastes as so much filth which must be got out of sight and mind as secretly and completely as possible. The result is that many who come to live in a cottage think only of installing a town-type water closet with an outlet, preferably to a main sewer or to a small-scale sewage system (cesspit or septic tank) with all the expense and waste of space this involves.

Such people have to extricate themselves from their early conditioning and learn afresh that human excrement or shit is nothing more than the unusable part of quite wholesome substances taken into the digestive system as food, processed and ejected by it as waste organic material. This waste when naturally decomposed may be brought back into the cycle of life as plant nutrients in the same way as other animal dung. It is only when the organism is diseased that excreta may carry infection through rats and flies. The simplest and most satisfactory method of disposal is to add this waste to the compost heap and cover it with a shovelful of earth to keep off the flies, when it will rapidly lose its noisome qualities and aid the heating of the other composted matter. In this way a valuable manure is conserved instead of being foolishly thrown away. Urine, too, contains nitrogens, phosphates and sulphates which can feed plants. The following disposal methods are set out in order of preference from the point of view of the conservation of the fertilizing properties of excrement.

An *earth closet*, placed just outside the back door or further afield, may have either a fixed or a movable receptacle with some means for the automatic strewing of dry, loamy earth after use. This may be held in a box over the seat, feeding through a hopper to a chute ending just above the bucket or tank at the back, with a flap to govern its flow; or it may be caused to fall down by the weight of the user on the seat. If a fixed tank is used, it should be one that can be emptied from the rear of the closet; if a bucket, the seat must lift up to let it be carried away and emptied. Either tank or bucket when new should be treated inside with a coat of tar, which will prolong its life and make it easier to empty. The work of emptying bucket or tank is the only drawback to this perfectly hygienic system.

Such work can be cut down a bit, and a concession made to city-bred prejudice by using instead a *chemical closet* such as the 'Elsan', which incorporates a ventilation chamber through which an air draught carries fumes to an outlet pipe behind the seat, and an inner container in which a chemical preparation, Elsan Fluid, diluted with water, is poured before it is brought into use. This quells the stink and reduces the bulk of the excreta without destroying its value as fertilizer. A rival chemical, Fertosan Odorcure, is a powder which, when sprinkled over the contents of the bucket, fosters the growth of minute fungi which feed on them and reduce their bulk. Both systems end with matter which can be used as an activator on compost heaps, and are viable so long as supplies of the chemical are assured.

There is now on the market, from Low Impact Technology, Wadebridge. Cornwall, a Swedish plastic-walled *latrine cum-compost heap* in which aerobic action is used to digest the excreta, which are dropped without either water or earth into a ventilated chamber with a sloping floor directly beneath the seat, where they slide down a 16-degree slope to a second chamber which is fitted with a lid for the tipping of kitchen and garden waste, and with perforated iron pipes to let in air. The mixture heats up and decomposes and then slides gradually under a partition into the third and largest chamber, from which it is removed and used on the garden. It is an ingenious and labour-saving system, but the cost and cumbersomeness of the structure must be weighed against the work it saves; for after all it does nothing which cannot be done just as well with an earth closet and a compost heap. However, it would seem preferable to a septic tank should the local authority accept it as an alternative, while insisting on either one or the other.

The same effect as the Swedish system at less cost can be gained by simple *composting in a privy pit*. For this one must have a transportable shed-type closet with door, seat and (for day-time lighting) a transparent vinyl roof. This is sited on a pair of reinforced concrete slabs laid over a fairly deep pit. Pit privies, properly made and maintained, are recognized as one of the safest sewage disposal methods. A pit vent can be installed, extending above the roof; the seat should be cut out of a board which can then be hinged backward to allow kitchen and garden waste to be thrown into the pit.

Having dug your pit and made sure that its base is above the ground-water level (place it so that runoff water will not flow through it), cover the bottom 2 ft with lawn mowings, leaves and straw. Over the top of the pit place the slabs, leaving a gap in the middle for fall of excreta and rubbish. Tip all household and garden garbage into the pit and add wood ash, cut grass and slaked lime with some earth once a week. When the contents are a foot or so from the surface, move slabs and closet to a new pit not less than 6 ft away. Level the first pit with 6 in. of grass and leaves and fill in with firmly rammed earth. When the second pit is filled the compost in the first will be ready for removal to the garden. Use a long-handled dung-fork to get it out. The pit can then be used again.

Both cesspools and septic tanks are wasteful and expensive methods of sewage disposal, but of the two the septic tank is much to be preferred. *Cesspools* have nothing to recommend them except that, like septic tanks, they permit the use of indoor W.C.s. Cesspools, which by regulation must not be of less than 4,000-gallon capacity as compared with the obligatory 600 gallons of a septic tank, are simply large storage containers in which sewage collects until it can be emptied, usually by the Local Authority with vacuum-operated tankers. What if this service breaks down? They become quite uneconomical when water from baths and sinks is run into them; and defective ones can be a source of soil contamination and water pollution.

The *septic tank* by contrast is really a miniature municipal sewage farm, more appropriate for the large country house than the humble cottage. All W.C., kitchen and bathroom wastes run through underground pipes to a distant brick-built settling tank on a concrete base. Here the solids in the waste are digested by anaerobic bacteria, during which process they give off methane or marsh gas into the atmosphere. When the fluid in this tank reaches a high enough level it floods over into a second, smaller tank from which it runs into a shallow filter-bed made of brickwork on a concrete base, and having a bed of clinker or coke. The effluent is spread over it either by perforated iron pipes or corrugated asbestos sheets placed at a slight slope and punctured with holes through which the fluid drips, and here aerobic activity breaks down the effluent to its final state, when it drips through an outlet pipe at the end of the system and is either discharged into a ditch or stream or made to run through land drains into the soil. The housewife should tip no disinfectant into such a system for fear of killing the bacteria on which its functioning depends.

If the settling tank is made airtight, and the methane gas it generates drawn off into a small-scale expanding gasometer, and from there piped to the kitchen, it can be used to fuel a stove of the Calor-gas type. In which case, insulation of the tank in winter is called for with maybe some gentle heating – in both cases to keep up the 85–90°F of warmth necessary for the multiplying bacteria.

Part II

*More Essential
Amenities*

CHAPTER SIX

WOOD FIRES AND STOVES

Quite often, old farmhouses and cottages have fairly modern fireplaces which, if removed, reveal that a Victorian iron range or suchlike has once stood there, the original open chimney piece having been bricked in to take it. These bricks can be knocked away to expose the original hearth and wide chimney of the pre-industrial hearth with its down-fire, its great breast-beam and chimney-corners. There is then a choice – to keep the original hearth and burn logs, or to instal a closed stove which will burn not only wood but peat, anthracite, coke and coal. Before deciding, survey the likely future fuel supplies in the neighbourhood. If there is an abundant supply of cheap or free wood, there is much to be said for an open hearth, even though much of the heat from it goes up the chimney.

There should be a large solid hearth-stone a few inches above the floor, an iron fireback and iron dogs across which to lay the logs. Logs may be three or four feet long and four to six inches thick, for the longer the logs your fire will take, the less labour there is in cutting them. One large, heavy log at the back of the fire, propped by stones or held by a chain to keep it from rolling into the flame, will produce a slow, steady heat for weeks before it is consumed. As the wood burns, leave the ashes to gather on the hearth-stone for a week or more. They will keep the heat all night. The fire will die right down to embers, but these have only to be scattered with shavings and light branches in the morning and gently breathed up with the bellows and they will spring to life, so that a good fire may be had very quickly. If a glowing stick or log is buried under a pile of ash in the hearth, next morning it will still be alight and can be blown into a flame.

If coal is to be burned on an open fire it is better to have a modern fireplace with a clay fireback which throws heat into the room. Never use coal on a wood fire, for the ash and clinker of even a few coals destroy the purity of a wood fire. Peat can be used, but if peat and wood are burned together a hard deposit may form in the chimney which is difficult to remove.

IRON STOVES. But you may not have such a hearth and chimney, and the kitchen and hall to go with them. You may have one large living-room which you want to heat by wood, as cheaply as possible. An enclosed iron stove, with a top opening for a metal chimney, may meet your needs. There is one called the 'Forester' which burns quite long logs and keeps in all night without refuelling. The old 'Tortoise' stove has its merits. Some of these iron stoves are very efficient, and will become red hot without danger or damage. They should be placed on a base of concrete, brick or stone.

If you want a modern stove which will cook your food and heat your bath and kitchen sink water and warm the room as well, the 'Aga' or 'Esse Century', which have solid metal fireboxes, or the smaller and cheaper 'Rayburn', which has fire-bricks, will meet your needs; but while the 'Aga' will burn only smokeless fuels (coke, anthracite and patent nuts), the others will burn coal, wood and peat as well. It is much cheaper to buy these stoves second-hand than new. One disadvantage of the 'Rayburn' is that its bottom fire-bricks burn out and crumble away with use and need to be replaced so it is as well to have a spare set, or half-set, stored away.

KINDLING A FIRE. In the absence of paper perhaps the best kindling for a fire is the finely shredded bark of the birch-tree, which is oily and burns with a smoky flame like a candle. It would be a thrifty practice to strip any birch-logs you have of their bark and keep this for kindling. From the garden, dried sunflower and artichoke stalks make good kindling. Fir cones can be collected in sacks by children, and used to start the fire. Among unusual fuels are the otherwise useless timbers from demolished buildings: when sawing these, beware of hidden nails which can badly damage the saw blade. The sea yields driftwood daily in rough weather, and in some areas sea-coal. For charcoal and peat, see Chapters 8 and 9. When lighting a peat fire, remember that it will not catch from paper only but needs a basis of wood kindling. Large unbroken turves should not be put on to a newly lighted fire – they will put it out. Instead, break them up between the fingers and feed the fire with them gradually, giving it your whole attention for the first ten minutes. When the fire has taken hold, pile the hearth up with large sods; if they appear to quell the blaze, it will soon shoot up again. A peat fire will call for more use of the bellows than a wood or coal fire. As peat is bulkier than coal in relation to the time it takes to burn it needs larger containers than the normal coal-scuttle. Though it requires more attention while burning, peat makes less dirt than coal and has a pleasant fragrance. Stack the peats closely at night and they will be found still glowing at the centre next morning.

CHAPTER SEVEN

GETTING WOOD FOR FUEL

The modern tool for felling trees and cutting up logs is the portable chain-saw. This is a ferocious implement consisting of a small petrol engine held within the frame of the saw and operating a band or chain with teeth which chew into the wood. The noise, weight and vibration are compensated for by the saw's capacity for cutting through logs like cheese, reducing a huge tree to manageable baulks of timber in an incredibly short time. Many farmers now have one of these things. If a large tree near your house is blown down and you wish to have the wood for fuel you could perhaps hire or borrow a chain-saw. It is hardly worth while buying one.

The ordinary tools required for wood-cutting in the open, felling trees, removing branches live and dead and cutting hedges back are: a tubular-frame coarse-toothed bow saw with a blade that is secured with a wing-nut; a hatchet, an axe and a billhook. You may also need a sledge-hammer and two or three iron wedges. For bringing down large trees a cross-cut saw worked by two men is needed; but the beginning woodman should attempt only trees of moderate size – say, not more than 30 ft high and a foot in diameter at the base.

FELLING A TREE. An experienced woodman will fell a large tree by making the cut as near to the ground as possible. The first stage is called laying in, when the buttresses or swellings leading to the roots are sheared off with an axe. This done, a tapering gap is made on the face of the trunk – the part facing the way the tree is to fall – taking away the support on that side. The axe is now put aside, and two men take up the cross-cut saw and begin to cut on the back of the tree. All the work of a cross-cut is done on the pull stroke, each man in turn pulling and releasing, or following-through the other's pull. Until a hundred years ago trees were normally felled with the axe alone. In that case, after the first opening was made on the face as described, the second opening on the back was made somewhat higher, causing the tree to go down more readily. Felling is always done in the winter months after Christmas. No one would wish to fell a tree when its sap was rising or it was in full leaf.

Suppose you wish to take down a tree of moderate size. You will notice first whether there is enough space for it to fall in the direction towards which it leans and towards which the height of its top branches inclines it to sway. Hew out a wide notch with the axe, on the back of the tree, and as low as is practicable. Putting the axe on one side, continue cutting with the cross-cut saw, but stop as soon as there are signs of movement in the tree. Take out the saw and tap one or two iron wedges into the cut, using a sledge-hammer or the heel of the axe. With a little skill one can direct in this way where the tree is to lie when it falls. A less deft way is to fasten a rope to the tree at about three-quarters of its height before cutting begins, the rope being long enough to pass well beyond the point where the top-spray of the tree will strike the ground. When the tree begins to topple, get your helpers to pull together in the right direction on the rope while you drive the wedges home. Once fallen – and do avoid letting the tree fall so that it leans propped up against another tree – the limbs can be sheared off with axe and saw and the trunk cut into movable sections with the cross-cut. Keep out of the way of suddenly falling limbs and the crushing weight of the trunk when it rolls off balance as a heavy limb is removed.

To REMOVE SIDE-BRANCHES from a growing tree so placed that it is not convenient to let them drop to the ground, dismember them piecemeal, sawing through branches one by one and lowering them by a rope. The rope, which should have a heavy iron hook spliced in to one end, may be of good manilla or nylon of $\frac{3}{4}$-in. diameter and in length rather more than twice the height of the uppermost bough to be taken off. Begin, though, with the lowest bough. Climb the tree, fix the rope round the branch at the point of balance by passing it round the branch and catching it in the hook. Pass the other free end of the rope over a branch or fork a few feet higher up the tree. Bring the free end of the rope to the ground and let a helper secure it by taking a turn of it round the trunk and hold on to the end, making sure there is no slackness in the operational part of the rope. Then begin sawing the branch with your bow saw, first cutting an inch or two upwards on the lower side, and then almost all the way through from above. If the rope is kept taut you need not fear the branch will suddenly break away. When the saw is through except for the last $\frac{1}{2}$ inch, get off the tree and gradually ease off the turn of the rope round the trunk. The limb will crack and come away, and may be safely lowered. If you have no helper then make the bight of the rope fast with a hitch while you are up the tree.

SAWING LOGS. To convert tree timber into fuel you will need a rough horse to hold the baulk of timber while it is being sawn. This is a strong, heavy frame of rough timber, made by constructing two wooden 'X's and joining them at about 3 ft apart by stout struts and crosspieces. The bottom part of the 'X' makes the feet while the top part holds the log to be sawn, the bottom part being larger than the top, for stability.

The work of sawing is halved if two people saw the logs, using the cross-cut. As with felling a tree the knack is to keep the saw running smoothly to and fro its whole length every stroke without jamming or bending – keep it perfectly straight. When sawing very large pieces, tap a wedge into the cut into which the saw is disappearing, knocking it a little further in as sawing proceeds. Keep an oil-can handy, as a few drops on the saw now and then will ease its way through resistant wood. After use, brush all sawdust out of the teeth of the cross-cut and rub it over with an oily rag.

CHOPPING BILLETS. If the sawn pieces are too big for the stove, they must be split on the chopping block, the stump of a tree about 3 ft high and 2 ft across standing solid on the ground. Up-end the billet on the block and bring down the axe with a sharp blow near its centre. It should split instantaneously. The straighter the grain of the wood, the less force is needed for the blow: straight-grained wood splits of itself, fibrous woods like fir and oak need more force, while cross-grained timber may defy the axe, yielding only to wedges driven in by repeated blows of the sledge.

Logs will stand the weather if stacked on their sides in the open, but will burn better if kept dry. They may be protected with tarpaulin, plastic sacks, corrugated iron or old lino. Billets and cleft wood need to be kept under cover in a fuel store; if there is room enough to house the tools and horse also, wood can be chopped and sawn on wet days.

FAGGOTS. The old woodmen wasted no part of the tree, and where the modern chain-saw man will too often burn the mountain of twigs left after his operations, leaving a charred circle in the soil, they would deftly slice off the twigs from the branches, lay them together in long bunches and bind them tightly with withies or wire, in this way making them into combustible faggots. Similarly, when hedges were cut, the wood was separated into three piles, for pea-sticks, for bean-poles, and the leavings for faggots. Faggots were burned whole in the tunnel of the cottage bread-oven. Chopped with the bill on the block into short lengths they can be used for kindling. They can be stored in the open in a large pile with the bundles laid lengthwise on the top and sloping like a roof.

Faggots can also be used for the floor of the calf or goat pen, with straw litter on top: the straw keeps cleaner this way as the animals' urine passes through it and drains away through the faggots.

TREE ROOTS. Stumps may be grubbed up by a tractor with a set of chains, or by a hand-operated monkey, an engine fitted with a low-geared winch worked by a lever which enables one man to pull the stump slowly out of the ground with some of its roots attached. To do this, attach the cable securely to the stump and fasten the other end to a stump or a standing tree-trunk. When working the winch you stand with your back to the stump to be pulled out. If you are faced with large numbers of tree-stumps, it may be quicker to blast them out with explosive. Tree-stumps, having twisted and distorted fibres, are difficult to split or saw and may be burned on the site.

SPLITTING A TREE-TRUNK. One sometimes comes up against a long, heavy trunk which can neither be moved nor easily sawn. Sometimes such trunks can be split open along the grain of the wood, using a heavy sledge, three or four iron wedges, and half a dozen larger beechwood wedges for keeping the gap open once it is made. This works better with ash, beech, oak and the softwoods than with elm. Use a crowbar for levering open the split at the last stage. Having split such a trunk you have two or more lengthy baulks of timber which can either be sawn into handy sections or hauled away as they are.

Most timber from full-grown trees is too valuable for fuel. Remember that your fallen tree might fetch a good price from a timber merchant before you begin cutting it to pieces. Wood sold for burning is usually branchwood or coppice stems, the heat potential of which depends on the amount of seasoning it has suffered. It is best to buy the winter's store in the previous spring and allow it to season in the open throughout the summer, preferably under an open shelter. When buying, too, remember that unseasoned wood, since it contains more moisture, is heavier than seasoned: so if the price is the same, buy the seasoned.

CHAPTER EIGHT

CHARCOAL

Before coal was extensively mined in Britain, charcoal for domestic and light industrial use was produced by the burning of logs at their place of growth, the forests, in such a way that the charred remnants could be the more easily carried away and stored for fuel. Charcoal is very light in weight, and yet it is almost the perfect fuel as it yields steady high temperatures without smoke or ash. 100 parts of wood yield about 60 parts by measure, or 25 parts by weight, of charcoal. So 25 cwt of green timber will give about 30 bushels, or 5 cwt, of charcoal. Its domestic use today is as a starter for anthracite stoves; but a charcoal stove can be used indoors not only to grill meat and fish but to bake scones, roast potatoes and boil vegetables and stews.

Charcoal is produced by heating wood in the absence of sufficient air for complete combustion. The heat first drives out the water from the wood, then the volatile compounds including creosote and tar, leaving the lightweight solids: black carbon with some mineral ash. Almost any wood will make charcoal, chestnut and coppice oak being considered best. No special tools are needed; the rough wood is just cut into lengths of from 2 to 4 ft and stacked for several months to season. Burning is done in the summer.

Charcoal is now made commercially in sealed metal furnaces from which the liquid by-products are run off for use. This suggests the possibility, which might bear investigation, of burning charcoal in discarded oil drums in which the draught could be easily controlled. Fluids could be left to drain off through holes drilled in the base of the drum, which would be raised above ground level on breeze-blocks, and piped or channelled into small tins or drums.

CHAPTER NINE

PEAT

Peat, or turf, has been the most widely used household fuel in Ireland and Scotland for centuries and is still the main fuel burned in the open fires or kitchen ranges in western Ireland homes. The right to cut peat on another's land or on common land is called right of turbary, 'turbary' being another name for bog peat.

Peat consists of partly rotted vegetable matter and not, like coal, of extinct varieties of lycopodiaceous trees compressed and chemically altered into hard carbon. The conditions for its formation are the presence of water and decaying vegetation and the scarcity of oxygen. It is said to take a hundred years to form a 6-ft depth of peat, and the depth of a bed may vary from two to twenty feet. The best slow-burning peat is coal black in colour, the poorer type a light brown. The time for cutting is in April or May, for it must have a full summer to dry out before it can be used the following winter. If carting is difficult it may be left until later when the bog surface is drier.

CUTTING PEAT. First, the top layer of grass and soil is stripped off to get at the peat bed itself, using a broad-bladed spade, with an old scythe blade or hay knife to cut into the rough herbage. Next the peat itself is cut, customarily with a special spade with a foot-long blade 6 in. across, along one edge of which another spur-shaped blade, the 'feather', is set at right angles, so that it will cut out an oblong block of peat in one operation. A block should be about 12 to 18 in. long and from 4 to 6 in. broad, by about the same thickness.

Peat cutting is best done with a gang, or a pair, one to cut, the others to stack, since wet peat is heavy. Cut at a vertical face extending through the peat bed to the mineral soil beneath, unless the bed is a very deep one. Take the face steadily backwards as the work goes on until the whole bed is worked out.

If the peats are not to be dried on the spot, they are carried to a drying ground and there spread out singly. They are turned after a few days so that both sides dry evenly. Later they are turned again and stacked on their ends, leaning together like little wigwams. This lets them dry out in the wind, this drying being completed later in larger, open-work stacks. As peats dry out they shrink and lose weight; a ton of fresh-cut peat makes only 3 or 4 cwt of dry fuel. When dry they are carted to the cottage and stored under cover or in close stacks with sloping sides to shed the rain.

HEATHER TURF OR SPINE TURF, as distinct from bog peat, is developed on drier heaths where heather grows freely – on Exmoor, the New Forest and the heaths in Dorset especially. Only 5 or 6 in. thick, it seems to be produced fairly rapidly, and in the New Forest the rule for cutting used to be to take one turf and leave two, on Exmoor to take alternate strips – this so that the heather on each side should overgrow the pared surface and renew the turf. Its relative dryness means that little preparation for burning is needed if it is cut in the summer.

CHAPTER TEN

OIL LAMPS AND CANDLES

Although convenient, electric lighting utterly destroys the poetic apprehension of the difference between night and day and resembles flame lighting about as much as the electric fire resembles the fire of wood, peat or coal. The old-time cottager lit his home in winter with tallow candles, or with rush-lights made by stripping all the skin except a narrow vertical ribbon from the pith of a meadow-rush and soaking it in a dish of melted fat. The dried rush-light was nipped between a pair of metal pincers on an iron stand and shifted forward from time to time as it burned down: Cobbett preferred these to candles, largely on the ground of cheapness.

Paraffin lamps later became popular because the oil was cheap, and because its volatile nature allowed it to ascend a wick by capillary action and to be present for burning at the wick top where it would vaporize fairly easily. Fish and vegetable oils were inferior in these respects. The lightest of these, colza or rape-seed oil, needed a lamp with a gravity-feed system with a separate raised reservoir and a complex of valves and vents to get the oil where it was wanted and to drain away overflow. Without such a machine, such oils can be burned only in the simplest lamps of all, the ancient cruses, with or without spouts and lids, where the burning point of the wick is in direct contact with the fuel, and which give only a feeble light. In the early years of the century the houses of the well-to-do were sometimes lit by acetylene gas produced by the action of water on carbide of calcium, which produced a brilliant white light. Carbide today is unobtainable in the shops, and will burn only in specially made, outmoded lamps. The same goes for air gas, a mixture of air with a small amount of petrol vapour.

This leaves us today with Calor gas, which can be fed

from large cylinders through a system of supply pipes, or from a small cylinder connected to a single lamp; it is effective, but is more economical for heating and cooking than for lighting.

In the absence of electricity the country dweller must depend either upon Calor gas or upon the *paraffin lamp* for night lighting. The most efficient hanging or table lamps are of the 'Aladdin' type, which have an incandescent mantle of asbestos fibre suspended on a wire frame above a circular wick, the whole being mounted upon the oil reservoir and enclosed by a glass chimney. The soft 100-watt light can be improved by slipping a second glass chimney over the first. If the wick is turned too high the mantle will slowly blacken with soot: rectify by turning the wick low so that the charcoal on mantle and chimney are burned away. If a little fine salt is sprinkled down the glass the mantle will clear itself more quickly.

The 'Tilley' type pressure lamp is really a safety storm-lantern. It burns with a hissing noise, and can be used indoors or out. It has a mantle which grips the top of a central burner stem up which the oil is forced by air-pressure from a pump set obliquely into the reservoir. Before it will light the mantle and stem have to be pre-heated with methylated spirit. When the oil reaches the burner at the top of the stem it is vaporized and turns into inflammable gas. The lamp slowly loses pressure and has to be pumped up; there is also some tendency for air to leak out at the points where stem and pump enter the reservoir. The owner of either type of lamp should have a reserve of spare mantles, chimneys, special wicks ('Aladdin'), and washers and springs ('Tilley'), none of which can be improvised.

The simplest, most reliable but least efficient oil lamp is the standard flat-wick burner, consisting of reservoir, single or double wick, winding mechanism and chimney. Its outdoor counterpart is the tinned metal storm lantern. Spare wicks and chimneys only are needed.

CANDLES give more light than simple cruses. The best are of beeswax, the next best of paraffin wax, the worst of tallow, which is any kind of hard animal fat. Tallow must first be rendered in boiling water, when the fat floats to the top and is skimmed off, leaving the cellular fabric behind. Beeswax may be bought from a beekeeper, paraffin wax from an ironmonger and oil shop. Candles can be made in moulds such as the casing of a bicycle pump sawn through lengthwise, or a disposable plastic liquid-soap container which, if used only once, need not be cut in two. A rustic mould is a length of dried cow-parsley stalk cut between the 'knuckles'. Insert a length of string for wick, and centre it by tying each end to the middle of a nail or matchstick resting upon top and bottom of the mould. Bind the two parts of the mould together and fix in an upright position. Pour in melted wax or salt-free tallow at the top end. Allow to harden, and add more wax as needed. Unbind the mould, dip it in hot water and take out the cold candle. Make and store a quantity of candles, as they harden and improve with age. In the absence of suitable moulds, make a cylindrical impression in a pile of wet sand, fasten the bottom end of the wick to a metal or stone weight. The thicker the candle the thicker in proportion must be the wick: an old bootlace makes a good wick for a thick candle. Dip all wicks in boracic to make them burn brighter. Lighted candles need to be trimmed now and then by the nipping off of a charred piece of the burning wick. As well as being pushed into a bottle or candlestick – first dipping the base of the candle in hot water – a candle may be spiked on an upturned nail.

Part III

Gardening for Survival

CHAPTER ELEVEN

THE VEGETABLE GARDEN

MAKING THE GARDEN. Sometimes the new tenant of a cottage inherits a well-made garden, but he is as likely to find a neglected area overgrown with rank weeds, a waste plot covered with brickbats, rusty stoves and bicycle frames, bottles and cans or a patch of virgin grassland. If he is fortunate the garden will have a southerly aspect, will be protected from winds and be unshaded by trees.

Grassland is easiest to deal with. Strip the turf from the soil with mattock or spade and pile it in heaps, grass side down, to rot into friable humus. If the garden is a rubbish dump, the bricks and large stones must be wheeled away and stacked in piles from which they can be retrieved for making paths or drain channels, while broken bed-frames, bottles and cans must be buried. (Some of these metal objects can be used later when reinforced concrete is wanted.) Metal containers may be crushed flat but glass bottles should be left whole and buried bottom-up in sand or ash in the pathways.

Neglected gardens or virgin land can be turned over to two or more pigs for cleaning and ploughing. The plot must be well fenced – a two-strand electric fence will do – and there must be clean water for the pigs to drink and simple huts for shelter. A pig house can be made from straw bales or from two sheets of corrugated iron fastened along the top and spread open at the bottom, or from thatched hurdles. If grass and weeds are high they should first be scythed, raked and stacked in a corner to rot. Brambles should be cut and burned. Scatter the pigs' rations, in the form of cubes, over the plot. In a year a couple of pigs will convert a half-acre wilderness into a ploughed, weeded and dunged area which needs only to be hand-harrowed to be ready for planting.

The question of gardening without digging will be dealt

with later, but it should be noted here that it is possible to start gardening on grassland by planting potatoes on the grass and then covering them with damp straw to the depth of 1 ft, when the grass will rot down to nutrient matter on which the potatoes will feed.* The straw can be moved the following season and a crop of peas and beans sown where the potatoes were.

Without pigs or straw, the garden will have to be ploughed or dug. Grass and annual weeds are no trouble, but perennial growth like couch grass, thistles and coltsfoot may be a nuisance. Farmers have found that couch grass (twitch) will disintegrate if ploughed under every month from February to August, when it will restore to the soil as much plant food as a crop of mustard grown for green manure. Thistles and coltsfoot can be banished in the same way, by ploughing or digging. Some obdurate weeds, like docks, will have to be grubbed out and thrown on the compost heap (q.v.), after being dried in the sun or crushed to pulp.

GARDEN TOOLS. With few exceptions the best tools are the traditional ones. Good tools may sometimes be picked up very cheaply at country auction sales, where they are trussed in assorted bundles and sold as 'outside effects'. Buy two or three such bundles, select the tools which suit you, tie up the rest and put them back for sale. If you buy new never let the price of a really good tool deter you. It is wasted money to buy inferior tools. Don't buy on sight but try out every tool for balance, and only when you are sure that it is right for you, put down your money.

You will need:

1. *Garden spade.* For rough digging and excavating. The spade should balance well and have a spring-tempered, all-steel blade (not stainless) with a tread on the top edge to protect your instep. Clean the blade after use with a piece of

---

* See W. E. Shewell-Cooper, *The Complete Vegetable Grower* (London: Faber, 1974) p. 87.

slate or shaped hardwood. Rub down all tool handles once a year with linseed oil. Wipe blades with an oily rag.

2. *Garden fork.* A 4-tined fork is needed for breaking up dug ground, lifting roots, etc. A flat-tined potato-lifting fork and a round-tined dung-fork are also useful.

3. *Mattock.* An invaluable tool, being a steel blade like a long, heavy hoe set at right angles to a handle of shaped ash. The blade needs to be about 9 in. long by 6 in. wide, and the handle stout in proportion. Use with a swinging movement for taking out trenches for potato planting, for slicing turf from new ground, for rough digging and heavy hoeing.

4. *Crome* (Fig. 10). A blacksmith can make one by bending the tines of a garden fork at right angles to the shaft and replacing the fork handle by a 5-ft ash pole. Used for mucking out sheds, this is a useful tool for breaking up the soil to a depth of 4 or 5 in. in 'no-digging' gardening.

5. *Hand hoe.* You need two hoes, a dutch hoe with a flat head that is pushed to cut down weeds just below soil level, and a draw hoe with a swan neck used to single out seedlings or to draw soil around the stems of plants.

6. *Garden rake.* Get one with a long handle, ten teeth and with teeth and neck forged in one piece. Don't use broomsticks as hoe or rake handles.

7. *Hand trowel.* For digging up small plants, potting, transplanting. If it is just 1 ft long it will serve for measuring distances between plants and rows. A hand fork is also useful.

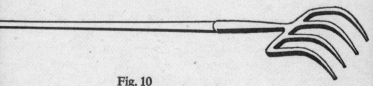

Fig. 10

8. *Hand harrow* or *cultivator.* Has 3 or 5 spear-point prongs curved like grapnels at the end of a long straight handle. Choose one with fixed prongs. (In general, avoid all

composite and multipurpose tools and those with inter-changeable parts.)

9. *Dibber*. A shortened fork-handle ending in a steel-shod point for making holes for leeks, potatoes, onion setts, cabbages.

10. *Wheelbarrow*. A big, rugged barrow with strong legs set wide apart, bought from a builders' merchant, for carting earth, muck and compost, potatoes, etc.

11. *Seed drill* and *garden line*. The line, tied at each end to an 18-in. stake, must be as long as the widest part of your garden plot. It is used as a guide to the draw hoe when it makes the drill for seed-sowing, the earth being filled in afterwards with the rake and firmed with the rake-head or by tramping. But the best way to sow seed is to use a seed-drill having a seed hopper with release mechanism geared to a wheel which runs along the ground and a variable arm which enables the user to keep the wheel at the required distance from the previous row.

12. *Other tools*. A 1½-gallon or 2-gallon watering-can with screwed-on rose. A plastic or rubber garden hose on a reel. A long-handled spade without the usual hand-grip reduces bending: used to sling earth from one spot to another. A roadman's shovel for sand and gravel, and a riddle for separating fine soil, etc., from coarse. Some shallow wooden boxes for seed-trays; flower pots; some rows of glass or plastic cloches. A caveat: Don't garden in Wellingtons, to which wet soil so readily adheres, but wear thick-soled leather boots.

PLANNING THE GARDEN. We shall not be concerned here with the ornamental but solely with the practical aspects of gardening, the growing of vegetables and fruits for food. Let us assume that your garden is south-facing or that it gets enough sunshine throughout the day, that it is not over-shaded, that the soil is reasonably fertile and easy to work, and that it is large enough to grow the vegetables you require throughout the year. How big is 'large enough'? It has been estimated that a garden of one-eighth of an acre (= half a rood = 604 sq. yards), i.e., one 60 yards long by 10 yards wide, or 30 yards long by 20 yards wide, will provide all the vegetables needed by a family of four, without the luxuries, and including potatoes; and that such a garden would take 288 man-hours a year to work properly – on average, less than an hour a day.* With a garden twice as large, and with less than twice as much work, one would be able to plant fruit bushes, some apple and pear trees and a strawberry bed; and with anything much larger there would be scope for growing a patch of cereals for home use and food for stock, or for sale.

Just as simple gardening can be done with only a spade, fork, rake and hoe, so it can be done without a planned garden, and without the shed, greenhouse, frame and hotbed that advanced gardeners find so necessary. But it is always best to think ahead from the beginning even with the roughest garden and the scantiest equipment, and to plan the garden for the easy working of a four-course rotation system by dividing it into four main plots for this purpose, with a further plot for such permanent crops as rhubarb, asparagus, cucumbers, vegetable marrows and perhaps Jerusalem artichokes (Table 1).

Table I

ROTATION OF CROPS

Plot 1	Plot 2	Plot 3	Plot 4
Beans, Peas, Celery	Beet, Carrots, Parsnips	Cabbage, Broccoli, Savoy, Sprouts, Cauliflower, Turnips, Onions, Leeks, Shallots, Lettuces, Spinach	Potatoes
(move to next plot the following year)	→	→	→

Plot 5

Asparagus, Horseradish, Rhubarb, Seakale, Herbs, Artichokes, Mustard and Cress, etc. Bush fruits: Currants, Gooseberries, Raspberries. Strawberry bed (move every 3 years). Seed bed for raising brassicas.

* Shewell-Cooper, op. cit., p. 20.

Assuming the garden to be rectangular, divide it into five plots of a size to be governed by the quantities of the various vegetables it is intended to grow. The fifth section is set aside for perennial and secondary crops, the other four for crop rotation, the plan repeating itself every four years.

The purpose of rotation is to give the land three years to recover from the food demands of a particular type of crop while allowing that crop in the following year to avoid whatever pests or disease spores peculiar to itself have been left behind in the soil, and by which the successor crop in the rotation will be unaffected. At the same time it allows a system of manuring to be carried out, a large quantity of organic manure always being given to the plot planted with potatoes, and slaked lime at the rate of 1–2 lb to the square yard to the next plot before it is sown with legumes. The legumes then hand over to the following crop the nitrogen they have fixed in their roots.

A four-plot garden, with a further, fixed plot at the far end, is obviously best approached from a path down the middle, with perhaps another running across at right angles. Any lawn will be near the house, perhaps separated from it by a paved area serving in summer as an outside room. Place the herb garden near the kitchen door. Toolshed, potting and storage sheds, if you have them, should be concealed by a hedge or by a row of sunflowers or artichokes at the near end of the garden, as should the cold frame and greenhouse, if any. Cultivate right up to the edge of the garden, leaving no grassy, weedy verge. Make paths hard enough to take a loaded barrow without crumbling. Essential to the housewife will be a path to the clothes-line, so decide with her where that is to be fixed and make a path to it which will keep her feet dry at all seasons. Wherever the compost heap is to be it will need a strong, concrete track running up to it. Minor paths may be of gravel or cinders or concrete slabs, and a path across a lawn is best made of spaced stepping-stones sunk level with the ground so that they do not obstruct scythe or mower.

FRUIT. If the garden is large enough, grow fruit on a permanent site at its far end. If there is room for the planting of a small orchard, grass it down and pigs, geese or hens can be run on it without harm to the trees. The beginner will find that less skill in pruning and training is required to grow apples and pears as bushes or standard or half-standard trees than as espaliers, fans, etc. Make sure that if the varieties chosen are not self-fertile they are paired with the right cross-pollinating variety, or they may not fruit. Always give new trees enough space for later growth. Soft fruit is best treated as a unit and grown where it can later be caged in a wire-netting enclosure to keep out birds, weeds inside the cage being suppressed by mulching. Loganberries and blackberries, however, are best grown outside, trained against fences or walls.

SIMPLE VEGETABLE GROWING. Most gardening books give equal emphasis to every vegetable, failing to point out to the tyro that there are degrees of difficulty in their cultivation. Some vegetables may be grown without being first raised in a heated greenhouse in special seed compost, and some may not; some are more or less immune from attack by particular pests and diseases, and others have to be nursed along and protected by spraying. The man with a new garden, and without those aids to advanced gardening, the greenhouse and the cold frame, should know that by choosing the easiest vegetables to grow he can begin straight away to provide food for his family through the year; although if the ground is very rough he may have to be content the first year with a cleansing crop of potatoes only, which will prepare the soil for mixed crops the next year.

The following table (Table II) will be found useful. Begin with the easiest vegetables to grow, leaving the more difficult ones until you have gained experience and the garden is well under control.

Table II

EASY AND DIFFICULT VEGETABLES

Group One: Very easy vegetables. These are relatively free from particular pests and diseases, are uncomplicated in cultivation, and need no glass or heat in their first stages.
Salads: Radish, all kinds of lettuce, spring onion.
Roots: Jerusalem artichoke, beetroot, parsnip, 1st and 2nd early potato, salsify, scorzonera, swede.
Stems and bulbs: Garlic, kohlrabi, leek, onion (from setts), rhubarb, shallot.
Legumes: Broad bean, haricot bean. (Pick off tops of broad beans when plants have reached full growth to deter blackfly.)
Leaves: All kinds of cabbage and kale, purple sprouting broccoli, spinach-beet, all herbs.

Group Two: Less easy vegetables.
Roots: Carrot (guard against fly), maincrop potato (spray against blight), turnip (guard against flea-beetle).
Stems and bulbs: Onion from seed (guard against fly).
Legumes: Pea, early and maincrop, runner bean (all need to be staked).
Leaves: Brussels sprout, spinach, cauliflower.
Various: Vegetable marrow, globe artichoke, pumpkin, strawberry, watercress.

Group Three: More difficult vegetables. These need to be raised under glass in special compost, are selective in their soil needs and require special treatment in cultivation.
Celery, chicory, ridge cucumber, squash, sweet corn. Asparagus, outdoor tomato. Grown under glass throughout: Aubergine, melon, cucumber, indoor tomato.

The gardener who for the first year or two grows only the vegetables in group one will be able to sow nearly all his seeds in the rows where the plants are to grow – only the cabbages and kales will need a special seed-bed of finely raked soil in the open, from which the seedlings will need to be transplanted.

(If, however, he particularly wants to raise one or two plants of group three he can do so by starting them in a box or tray of seed compost in any window of the house so long as the room has at least six hours of direct sunlight each day and is at a constant warm temperature. Place a layer of easily drained soil in the bottom of the tray and cover this with a ¾-in. layer of washed fine sand or shredded sphag-

num moss, pressing it down to make a firm, smooth seedbed.
Draw some $\frac{1}{2}$-in. furrows in this, then water thoroughly and
allow to drain. Sow seeds sparsely in the rows and cover
lightly with more sand or moss. Sprinkle lightly with water
and cover the tray with a sheet of clear plastic film, which
will hold moisture in the soil and air around the seeds. Now
place the covered tray in the window, where it will need no
further attention until the seedlings have grown their first
true leaves. They can then be transplanted into pots and
placed in a hotbed, a cold frame or under cloches until they
are ready to be hardened off and planted in the open.)

SOME GARDENING PROCEDURES. Information as to plant
varieties, sowing and planting out distances, special cultivation and season of growth, is to be found in any good
gardening handbook or seedsman's catalogue. Every gardener should have a reliable handbook with plenty of line
illustrations in the text, supplemented by a copy of – I sug-
gest – *The Complete Vegetable Grower* by W. E. Shewell-
Cooper and, as he gains experience, Lawrence D. Hills's
Grow Your Own Fruit and Vegetables. Both these writers
emphasize the need for the conservation of soil fertility and
the production of healthy fruit and vegetables by the use of
organic manures and home-made compost rather than chemical fertilizers. There is no doubt that organic gardening is
not only biologically sounder than gardening with chemicals but is cheaper and more self-sufficient. The making of
compost is a central activity in good gardening, for compost
not only provides all the plant foods found in farmyard
manure, it improves soil texture by encouraging the presence of earthworms, it holds moisture, and conserves heat
in the soil which is so important for early crops. Compost-
grown vegetables are in general healthier and tastier, if not
always larger, than vegetables grown without it. More, there
is an inherent satisfaction in composting in that it turns
materials which would otherwise be wasted to good account,
feeding them back into the natural cycle of life.

MAKING COMPOST. Compost can be made in a brick, metal

or wooden bin, or two or more bins placed side by side, each not less than 4 ft square, for smaller bins will not heat up or retain their heat so well.

The simplest bin is made by driving four metal stakes into the ground and fastening round these a length of 4-ft fence netting to make a square basket which can then be lined with hardboard, plywood or tough cardboard. A more satisfactory bin is made with $\frac{1}{2}$-in. boards sawn to length and nailed to four upright 3 in. $\times$ 3 in. corner posts on three sides. Leave the front open to receive slats cut to an inch or two less than 4 ft, and drop these in from the top between vertical runners as the heap is built up. The bin must be based on earth, and a horizontal air-channel made at the base by parallel bricks placed on edge, the end bricks protruding beneath the bottom front slat. Brush some preservative other than creosote over the wood, and follow with a coat of paint to prevent rotting. When the box is ready, start the heap at the bottom with some coarse woody material and continue by throwing in any organic wastes that will decompose: lawn mowings, hedge clippings, clean and diseased vegetable leaves and stems, kitchen refuse, waste hay and straw, soaked newspapers, rotten fruit, etc. The dry compost should be wetted throughout until it holds no more moisture than a squeezed-out sponge.

With two boxes you can throw material at random into the first box, and when it is full, use it for building a heap more carefully in the second box. The vegetable material needs animal matter as an 'activator' to start off and maintain the process of microbial fermentation which produces heat. All animal manures, including human, and particularly human urine – which should be diluted before sprinkling on – are activators and may be added in the proportion of 1 part animal to 4 parts vegetable matter. To build the heap methodically in the second box, then, begin with woody material, then add a 9-in. layer of vegetable matter from the top of box No. 1, scatter on this a thin layer of animal dung, follow with another thick layer of vegetation, sprinkle this with slaked lime and a spadeful of earth, add more vegetable matter, then more dung, and so on until the box or bin is full. Let the heap sink, and in a few days add more stuff in the same way.

Build the heap around a vertical air-channel made with a length of piping which can afterwards be withdrawn. Be sure to keep the heap moist; cover with sacks or old matting to retain heat, and in wet weather fit a rough roof of corrugated iron or of planks covered with roofing felt. After about 6 months, when fermentation is over and the compost has cooled, it should contain worms which have crawled into it from the soil beneath. It can then be shovelled into a wheel-barrow from ground level if two or more of the top retaining slats are removed and the bottom ones pushed upwards and wedged in place.

COMPOST FROM HENS. Cottagers who keep a few hens in a wired run with access to pasture can produce compost by littering the run with straw, bracken and garden weeds and scattering the birds' grain on this. In scratching, the birds break up the litter which is then mixed by their feet with their droppings. If litter is constantly added the surface of the run is kept fresh and clean, while as the litter deepens its lower layers are turned to compost on the spot. Clean out the run once a year, wet the material and stack it in a pile to generate heat and complete the composting. Each bird will produce in this way up to $\frac{1}{2}$ ton of compost a year.

ORGANIC MANURES. Strawy stable and farmyard manure, and horse manure that has been used for mushroom growing, should be bought in when possible. It should be mixed with compost, as should any goat and rabbit manure. Poultry droppings and pigeon dung are both useful activators for the compost heap, but are too fierce to use directly on the soil without being stored first for a year in a dry bin. Wood ash should be kept dry until wanted for use on the onion bed, and soot, too, should be stored dry for a year before using. Leaf-mould makes a good peat-like mulch; small amounts may be put on the compost heap, but very large quantities of leaves should be tipped in the autumn into a wire cage where they will stay put and rot down for use as mulch next year. Seaweed is rich in minerals, but not in humus; it is a compost activator. Some of our coastal sea sands contain calcareous and animal matters which are of benefit to soils and plants. Marl, a species of muddy, clayey subsoil containing calcium carbonate and other elements, can be used for soil improvement, while some districts possess deposits of volcanic subsoil which can be dug up and spread on soils with good effect. Lawrence D. Hills advocates growing Giant Russian Comfrey for compost and for laying fresh in the trench where potatoes are to be grown. Commercial organic fertilizers which can be applied direct or used in compost heaps are: dried blood, hoof and horn meal, wool shoddy, bone meal and steamed bone flour, and fish manure. Green manuring is a quick method of making humus and supplying nitrogen to the soil by first sowing broadcast, and then shallowly digging in on the ground where it grows, a crop such as rape, mustard, lupin, buckweat or winter tares, or a mixture such as clover and rye-grass; it is dug or disc-harrowed in while still lush, before flowering.

MINIMUM CULTIVATION. Nearly all old gardening books begin with a chapter stressing the need for deep digging of gardens every year, but modern gardeners are beginning to discover that this is by no means necessary. It takes a man 30 days to double-dig an acre of heavy land, 18 days to dig it one spit deep. Why go to great lengths to aerate and break up the soil (and then compact it again with boots and rollers) when in an average healthy soil there are millions of workers, the worms, already doing it for you? Digging deeply and burying top growth to a depth where it cannot easily rot into humus, it is now widely believed, breaks up soil structure and interferes with the capillary action by which moisture at a depth can percolate upwards to nourish the roots of plants in a dry season. The school of 'no-diggers' points out that unaided nature manages to produce forests of enormous trees without any form of soil cultivation; so why should man dig and plough in order to grow his corn and potatoes?

The answer seems to be that digging and ploughing are initially necessary to make a clearing in wild nature and to suppress unwanted growths. It is easier to sow and cover seeds in a loose, friable soil than in a compact one. It is evident that every modern no-digger must have begun with a piece of land that at some time has been under cultivation. In practice the success of the method seems to depend on the large scale manufacture of compost, of which there must be enough to make a permanent 1-in. to 2-in. layer over the whole garden. To manufacture this may be as much work as digging the garden! As well as compost, then, mulches of leaf-mould straw and even sawdust are used. Sawdust takes nitrogen from the soil, and if used (with discrimination) must be seasoned for a year by being spread on paths before being thrown on to the garden. Where compost is short, Shewell-Cooper for one recommends the use of sedge (not sphagnum) peat, the seeds being sown on the soil surface and then covered with a 1-in. layer of damp peat in which a fishmeal fertilizer has been mixed. Brassicas, etc., are dibbled into the peat in such a way that their roots just penetrate the soil beneath. Fresh supplies of peat are needed yearly to make good the quantity converted into humus by worms.

How MUCH TO GROW. Having marked out the plots for your four-course rotation and opted for the easy vegetables in group one with one or two from group two, a cropping plan is needed which will ensure a year-round supply of vegetables, not forgetting the hungry gap of late winter and early spring. For a family of four or five persons the following table (Table III) may be taken as a guide.

In such a garden plan there will still be room for successive sowings of lettuce, radish and spring onion through the summer. (If the garden is too small to grow all these crops, omit the 11 rows of maincrop potato and buy cheap potatoes for winter use in bulk from a farmer.) Such a plan

Table III

CROPPING PLAN FOR GARDEN OF ABOUT 500 SQ. YARDS PROVIDING VEGETABLES FOR A FAMILY OF 4–5 PERSONS

Vegetable	Quantity needed	No. of 30 ft rows needed	Quantity of seed or no. of plants required
Jerusalem artichoke	1 cwt	4	14 lb
Potato, 1st and 2nd early	3 cwt	8	32 lb
Potato, maincrop	5 cwt	11	33 lb
Cabbages of all kinds*	200 heads	10	200 plants
Kale	40 lb	2	40 plants
Kohlrabi	10 lb	1	⅛ oz
Leek	72 lb	4	240 plants
Onion	2 cwt	6	6 lb setts or ¾ oz seed
Shallot	42 lb	2	4 lb bulbs
Swede	1 cwt	4	1 oz
Beetroot*	56 lb	2	1 oz
Parsnip	56 lb	2	½ oz
Carrot*	84 lb	4	1 oz
Broad bean*	120 lb	4	1 pint
Runner bean	140 lb	4	1 pint
Pea*	1 cwt	4	2 pints

* *Successive sowings required*

provides for summer vegetables from June onward, when kohlrabi and the first broad beans appear, an abundance of late summer and autumn produce, roots and potatoes for winter storage, with leeks, parsnips and artichokes still to be dug in the garden up to April, and with cabbages for cutting right through from June to the following May.

STORING VEGETABLES. Vegetables that can be left to stand outside for winter use include: all kales, brussels sprouts, savoy cabbages, parsnips (which can be dug and heaped in criss-cross piles and left to get well frosted), swedes, leeks, artichokes. Vegetables that stand the winter to make growth in spring include: some lettuce, spring cabbage plants, and, as autumn-sown seeds, onions, peas and broad beans. All other vegetables must be harvested before the frosts and those not needed until the spring (part of the potato crop, beet, carrots, marrows) kept in an outdoor pit away from the house, while vegetables for winter use need to be near at hand in store-room or cellar.

The temperature in a store-room should be even, the atmosphere moist (especially for roots) and ventilation free. A dirt floor will keep air moist; with a cement floor, sand, soil or moss should be used in packing vegetables. Where no store cellar exists a small one can be dug outdoors in the side of a deep bank, or a 6 ft 6 in. high cellar dug underground, lined with concrete blocks and roofed with a flat reinforced concrete slab. Walls and roof must be strong enough to hold the weight of earth shovelled back over the top. No windows are needed but there must be a normal door approached down a flight of concrete steps. Litter the floor with peat and over this place a false floor of wood slats. For ventilation, build an air vent into one wall leading into a pipe made to protrude through the ground, its top screened to keep out mice and birds.

TREATMENT OF VEGETABLES FOR STORING. Twist off the green tops of beetroots and turnips and pack sound roots in boxes with dry sand or wood ash to prevent shrivelling or frost damage. Cut marrows and squashes before the first frost, leaving 2 in. of stem, and hang them in nets. Twist or plait the stems of onions round a central string, or pack into old nylon stockings and hang up on the wall. Stuff French and haricot bean plants whole into a large sack until it is full; tie at the neck and hang it from a nail while you beat it back and front with a stick until it is quite flat. Tip out the beans from the bottom of the sack, blow off the dust with bellows, and store the beans in large tins.

STORING ROOTS IN CLAMPS. Large quantities of potatoes, or roots for stock feed may be stored outdoors in piles covered with straw and earth. Lift the maincrop potatoes on a dry day when the haulms are ripe and the skins on the tubers are set. Select a patch of ground away from the drip of trees and not likely to be flooded in winter. Lay straw or ashes on the ground to a depth of 2 in. and arrange the tubers on this in a conical heap. Cover with a 6-in. thickness of straw and dig a drainage ditch near to it, covering the straw with earth taken from the ditch. Leave a tuft of straw to stick up through a vent hole at the top, but seal this with earth before heavy frosts come. Mangolds, fodder-beet, etc., are stored in the same way.

To STORE CABBAGES OUTDOORS, pull them out root and all and set them side by side with their roots in a shallow trench, packing loose soil around the roots. Then build a rough frame about 2 ft high around the trench with boards or poles, and bank this with soil. Place poles across the top of the frame to hold a covering of straw or bracken.

STORING FRUIT. Late varieties of pears and apples will usually store well, but allow the picked fruit to cool and sweat overnight before stowing them away in a cool, dark, well-aired place protected from mice and rats. Store fruit separately from vegetables. If stored in boxes, wrap each fruit individually in newspaper; if on shelves, do not allow fruit to touch one another. Avoid upstairs rooms or dry places which will shrivel the fruit. Inspect once a week and remove all rotten fruits.

SAVING SEED. In normal times it is better to buy seed than to save your own, but it sometimes happens that runner bean seed, for example, is in short supply, and those who have wisely let some of their bean pods dry out on the haulm for the sake of their ripe seeds will not be affected by the general scarcity. With potatoes it is thrifty to save each autumn one half of your seed needs for the following year from the most productive plants in the garden, buying the other half as new Scotch-grown tubers and then saving half of the following year's seed needs from the best plants produced by these, and so on from year to year. Again, with packet seeds it is advisable to buy new seeds each spring, for the life-span of seeds varies, and unless you are certain of the age of seeds and know their life-span (1 year for parsnips to 7 years for marrows), it may prove a waste of time to sow them. If, however, you do decide one year to save some of your own seeds, remember to distinguish between self-pollinating vegetables which reproduce themselves almost exactly through their seeds, and those which cross-pollinate and cannot be relied upon to breed true to type unless they are carefully isolated during growth. Peas, beans, tomatoes, lettuce are normally self-pollinated; cabbage, broccoli, turnips, cucumbers, onions, parsnips, beet, marrows are cross-pollinators. If seed is to be saved from one or two selected plants in the second group they should be protected from unwanted cross-pollination by tying the flower heads in a muslin or polythene bag. Plants that have been let run to seed are hung head downwards in a dry shed with newspaper spread on the floor to catch the seeds as the plant dries out and they fall. Store them in lidded jars or tins.

Part IV

Bread, Beer and Tobacco

CHAPTER TWELVE

FROM SEED CORN TO BREAD LOAF

The farm labourer of sixty years ago who had a half acre or more of garden or allotment with a pigsty at one end, would sometimes divide the area into two and grow wheat and vegetables alternately on each part, having prepared the ground by spreading the pig manure and digging it in in the autumn. He would sow the wheat broadcast or use a 3-ft dibber in each hand and a fixed central line, walking backwards around and about the line in a widening 'square circle' – the pattern of ploughing – followed by a woman or child to drop the seeds into the dibber holes; or he would use a hand drill. He would harvest the ripe crop with a sickle or a swap-hook and beat out the corn from the husk with a flail – implements which had been superseded on the farm but which survived in the farm-worker's yard.* From his quarter acre he would get a yield of perhaps 10 bushels of wheat, which he would get the miller to grind into household flour for bread and puddings, the coarse offal being fed to the pig.

Anyone who has a half-acre garden can still do the same, except that he will now have to grind his flour at home. The seed needed would be 2½ pecks. Since it takes a pound of flour to make a 1¼-lb loaf and 10 bushels of corn = 600 lb, allowing for wastage there would still be enough flour to make at least 1½ loaves daily throughout the year. And the very best sort of loaves, longer keeping, tastier and more nourishing than any bought bread.

Nor need one grow wheat only. Rye and barley make good bread, and may be used whole or mixed with wheat flour. In southern parts of England, maize also may be

* See George Ewart Evans, *The Farm and the Village* (London: Faber, 1969).

grown, ground into flour, and mixed with wheat flour to make bread.

GROWING WHEAT. Wheat is simple to grow but it needs heavy soil or good loam. Barley requires a light loam, while rye will grow on lighter soil. Different varieties of wheat suit different localities and soil types – so consult local seedsmen, or obtain seed from a neighbouring farmer who knows his oats.

Spring wheat needs sowing before the end of March, after which time it is better to sow barley. Deep cultivation of the soil is not necessary; as with other crops, there should be plenty of moisture-holding organic matter in the soil surface.

Sow with a hand-drill set for 4 in. between rows, taking a line in the centre of the plot and following in a 'square circle' round and round it to save pauses at the end of each row. Rake the plot, and roll it if you like. Take steps to scare away birds while the wheat is growing. Deal with large weeds like thistles and docks in June/July; smaller weeds will be suppressed by the wheat.

CUT THE WHEAT as soon as the grain is hard enough to rub out of the ear when rolled between the hands. The traditional harvesting tool is not the scythe but the serrated-edged sickle (Fig. 11). Such sickles are still made by T. & J. Hutton & Co. Ltd, Ridgeway, Sheffield, among others. A practised scythesman can work, it is true, about twice as fast as a man or woman with a sickle, cutting two acres to the sickle user's one. Barley and rye are more quickly cut than wheat. With a two-handed scythe you cut the corn-stalks close to the ground, letting them fall in a long swath. With the one-handed sickle the action is different. Bending over, you grasp a bunch of ears with one hand, thrust in the sickle with the other and draw it towards you with a gentle sawing motion, severing the stalks with the least spillage of grain. It is usual to cut near ground level, to bind the stalks and ears into sheaves and lean them together in stooks; but with a

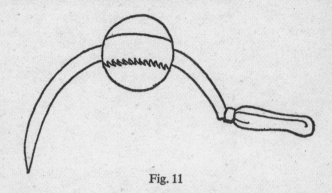

Fig. 11

sickle you can cut as high as you please, merely taking off the ears if you prefer. But as wheat straw is valuable it is best to follow normal practice, and thresh out the grain from the full stems, which can be afterwards stacked and used. Unlike the scythe or the smooth swaphook, the sickle needs no sharpening during the period of use.

Harvesting goes well if two people work together, one cutting the corn, the other gathering and tying the sheaves. A sheaf is a full double armful of stems clasped to the body and tied by touch at the far side with straw rope or binder twine. If full of weeds, tie sheaves smaller and leave to stand until the weeds have withered. Do not stack the sheaves until grain is hard with the straw all dead. Wet or weedy corn may generate heat when stacked, and spoil. Stacking is best done under an open-sided shed with a hard floor, or in the open on a square of polythene and with a black polythene or tarpaulin cover. The object of stacking is to store the wheat until such time as the grain can be thrashed out from the straw, and the sooner this is done the better.

To THRASH, untie the sheaves singly or in pairs, and spread the corn, indoors on a clean flat floor or outdoors on a canvas tilt, and whack out the grain from the ears either with a single stick, or with a double, jointed stick called a flail (Fig. 12), which can be made by taking a stout ash stick (the handle) and an equally stout stick of tough holly or blackthorn half its length (the 'swingel' or 'souple'), and jointing them with a piece of pigskin or eelskin or other tough flexible material in such a way that it forms a universal joint, allowing movement of the swingel in any direction.

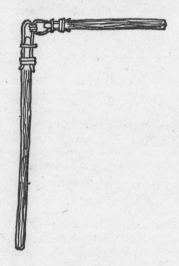

Fig. 12

The flail needs practice in its use, but when mastered does its work well. Having knocked out all the grain from the sheaves against the thrashing-floor, stack the straw and winnow the grain – i.e. free it from husks, bits of straw, weed seeds and dust – by tossing it in a draught which is just strong enough to blow away the trash without disturbing the grain. Where there is current, an electric fan or vacuum cleaner can be brought into service. The winnowed grain must then be stored in lidded bins until wanted for grinding. The labour of flailing can be saved by investing in a miniature grain thresher which can be clamped on to a workbench and worked by hand or a small electric engine. R. G. Garvie & Sons, of Aberdeen, make both hand threshers and hand winnowers which can be adapted for use with different cereals. A hand maize sheller is made by Ransomes Sims & Jefferies Ltd, Nacton Road, Ipswich.

GRINDING CORN. When there is nothing else to use one can grind corn in a fixed, hollow stone or 'saddle' using a flat moving stone or 'muller', the most primitive of all grinding devices. The quern, first used about 2,000 years ago, was the first hand-mill proper, consisting of two flat circular stones, the lower of which was slightly dished while the upper one was pierced in the centre and turned on a wood or metal pin. The operator dropped grain with one hand into the hole and with the other revolved the upper stone by means of a stick pushed into a small hole near the edge. Such stones, if found on ancient sites, and fitted together, can still be used. It is no longer possible to go to an ironmonger and buy a simple hand mill; but a small mill, weighing 15 lb, called the 'Junior', is made by Ets Champenois SA, Cousances-les-Forges, Meuse 55, France. For kitchen use, a coffee mill, or a mincer fitted with a corn grinding plate will grind small quantities. Some electric mixers like the 'Kenwood' have an attachment which will grind wheat; and a small electric mill with a hopper of 0·1 bushel capacity is marketed by 'CeCoCo' of Osaka, Japan. The flour ground will be wholemeal including the bran; the finer flour is obtained by sieving.

Baking

WHOLEMEAL BREAD. To make standard wholemeal loaves, take 3 lb flour, 1 tablespoon black treacle or 1 flat tablespoon soft brown sugar, 1 flat tablespoon salt, 1 oz fat (lard, margarine or butter), $1\frac{1}{2}$ pints lukewarm water, 2 oz fresh baker's yeast or 1 oz dried yeast. Warm a large crock (most mixing-bowls are too small), or a preserving pan. Then:

1. Put flour in crock or pan and stand it in a warm place.

2. Cream the yeast in a jug with 1 teaspoon of the sugar or treacle; add $\frac{1}{2}$ pint of the warm water, mix well and stand in a warm place until the yeast is working – about 5 to 10 minutes. Do not allow to get hot or the yeast will be killed.

3. Rub the fat into the warmed flour.

4. Dissolve salt and sugar or treacle in the remaining pint of warm water.

5. Make a well in the flour and pour in the salted water together with the yeast water.

6. Mix with wooden spoon, then knead well in the crock: there is no necessity to turn out on to a floured board to knead the dough. If the dough sticks to the side of the crock or pan, simply lift it away from the bottom and sprinkle a handful of extra flour into the pan, as often as required. Knead only for a few minutes until the dough is smooth and neither too wet nor too dry. Cover bottom of pan with another thin layer of flour, on which the dough should rest; then leave the pan (or crock), covered with a damp cloth, in a warm place until the dough has risen to twice its size or more.

7. Meanwhile, grease three 2-lb bread tins and put them to warm.

8. When the dough has risen, knead lightly again, then divide into three and place in the tins. There should be only so much dough in each tin to come a little less than half-way up the sides. If preferred, two 2-lb tins and one 1-lb tin can be used, the remaining dough being made into ten or twelve rolls each the size of a small apple.

9. Set the tins once more in a warm place until the dough has barely risen to the tops of the tins and the rolls, if any, are twice their original size.

10. Bake in a hot oven pre-heated to about 375°F for 30 or 35 minutes for loaves, 25 or 30 minutes for rolls.

11. Turn out on to a wire tray to cool.

TYPES OF OVEN. The quantities given in the above recipe are suitable for the housewife with a 'Rayburn' or similar stove which is kept burning all the time. Our forebears who baked only once a week or even less frequently, and made large quantities of bread at a time, did so to economize on fuel. The brick bread oven still remaining in some old cottages is a long tunnel, without grating or chimney, built into the wall, and is pre-heated by a fierce fire of faggots. Having lit the faggots you close the door, and when the fire is burnt out, take off the door and either rake out the ashes or brush them to one side before putting in the shaped dough and any other food which is to be cooked. Cooking is done wholly by the heat retained in the bricks. A 'tin' loaf is of course one baked in a tin, and a 'cottage' loaf one baked on the stone floor of the oven.

Before the coming of the late-Victorian iron range, cottages without brick ovens were supplied with a clay baking oven which retained heat in the same way as bricks after sticks had been burned in it. Its attainment of sufficient heat for baking is tested by scraping the top with a dry stick; if the stick sparks, it is time to take out the embers, wipe the oven out with a damp mop, push in the loaves, re-position the door and plaster it in place with clay. Such ovens are now rare.

If you have no oven but only an open hearth with cooking pots, then dough in tins can be baked in a big flat-topped fish-kettle over which hot embers can be piled. Time of baking varies with the heat of the oven and the size of the loaves but the oven should be hot (400°F) for the first ten minutes, preferably getting less hot, like the brick oven, as baking continues. At 375–400°F, a $1\frac{1}{4}$-lb loaf will bake in 45–60 minutes. Fresh yeast will rise quicker than old. Do not over-prove bread, put it in the oven when the dough has doubled its size. A loaf is cooked if it sounds hollow when tapped.

POTATO BREAD (a Victorian recipe). Get $\frac{1}{2}$ pint of yeast fresh from the brewhouse, weighing not less than $3\frac{1}{2}$ oz. Boil 1 lb potatoes in their skins. Place potatoes in a wood pail, add $\frac{1}{2}$ lb flour and pound them with the end of a rolling pin. Add 2 quarts cold water and mix well. Put 1 peck flour into a pan, make a well in the middle, place two sticks across the top, set a sieve over them and strain the liquor into the hole. Now add the $\frac{1}{2}$ pint yeast and stir in enough flour to make a thin batter. Cover, and leave for 2 hours to ferment. Add $1\frac{1}{2}$ oz salt and enough water to make the flour into dough. Knead. Cover with a cloth and let prove for $1\frac{1}{2}$ hours. Knead again and throw out on to the board, divide into loaves, prove, and bake in a moderate oven from 20 to 40 minutes.

BREAD WITHOUT KNEADING. For each 1 lb flour use $\frac{3}{4}$ pint warm water, $\frac{1}{2}$ oz yeast, $\frac{1}{4}$ oz salt. Mix yeast with a cup of the warm water and allow to work. Mix dry flour and salt in a bowl, make a well in the middle, pour the yeast into it and stir in from the sides to make a smooth paste. Slowly add the rest of the water and mix into a smooth wet dough, and let it rise in the bowl, covered with a damp cloth, until it doubles in size (20 minutes). Having got ready some greased tins, spoons baking powder, 1 teaspoon salt and about $\frac{1}{2}$ pint settle. Bake for 40–50 minutes.

BAKING POWDER BREAD. For every 1 lb flour, use 2 teaspoons baking power, 1 teaspoon salt and about $\frac{1}{2}$ pint milk. Well mix the flour, salt and baking powder. Moisten with milk to a stiffish dough, knead rapidly and divide into loaves. Place in greased tins and bake at 425°F for 1 hour.

MAIZE FLOUR AND WHEAT BREAD. Pour 4 quarts boiling water on 7 lb maize flour, stirring all the time. Let it stand until lukewarm, then mix into it 14 lb of fine wheat flour previously mixed with a $\frac{1}{4}$ lb salt. Hollow the surface of the mixture and pour into it 2 quarts yeast thickened to a cream with flour. Let it stand overnight. Next morning knead the mass and let stand for 3 hours. Divide into loaves and let them stand in tins for $\frac{1}{2}$ hour, then bake as with other breads.

BREAKFAST CEREALS. Almost any whole grain can be turned into a ready-to-serve breakfast cereal by simply exploding it in an iron or steel saucepan set over low heat for a few minutes with the lid on. Add a trace of salad oil and shake the pan now and then to prevent sticking. *Porridge:* Boil 1 pint water in a thick pan, add 1 level teaspoon salt (or mix salt beforehand with dry oatmeal), slowly sprinkle in 2 oz *medium* oatmeal, stirring continuously. When boiling again, reduce heat, put lid on pan and allow to simmer for 1 hour, stirring frequently (at least every 10 minutes). (Rather than a spoon use a 12-in. length of smooth $\frac{3}{4}$-in. dowelling with the end rounded, on which the cooking oatmeal cannot clog.) *Muesli:* With a small nut mill you can grind wheat, flax and sunflower seeds; add some milled nuts, dried or fresh fruit, and let the lot soak in milk. *Frumenty:* Put some ripe wheat into an earthenware jar, top with water, and let stand in an oven or on a stove overnight or until the grains burst. Eat hot or cold with milk and sugar.

CHAPTER THIRTEEN

GROWING AND CURING TOBACCO

The habit of smoking is one which it is better not to acquire, as it is injurious both to health and to the pocket. If acquired, the habit should be broken from time to time, for the state of dependence on a weed is not one which a moral being should permit in himself. One way of appeasing a craving for tobacco is to smoke a herbal mixture mainly consisting of dried coltsfoot, which grows along many country lanes. Proprietary herbal tobaccos almost all contain these ingredients: $3\frac{1}{2}$ oz coltsfoot mixed with $\frac{1}{2}$ oz each of eyebright, buckbean, betony, rosemary, wild thyme, lavender and camomile flowers. Using this recipe as a standard you can make up your own mixture to taste. While pleasant enough to smoke, the chief use of herbal tobaccos is as a remedy for asthma and other breathing troubles – to which end they need not be smoked at all but shaken loosely into a saucer and burned near the sufferer. Another tobacco substitute is the 'silks' of sweet corn.

How TO sow. Tobacco seeds are minute, $\frac{1}{2}$ oz of them providing plants for an acre of ground. As the Customs and Excise permit a man to grow no more than 25 lb of cured leaf a year, which is the amount of smoking tobacco yielded by 150 plants, you will require only a pinch of seed. Mix the seed with fine sand and sow in fine compost in a heated greenhouse in early March, or under cloches in early April. Prick out $1\frac{1}{2}$ in. apart when the first true leaves are an inch high. After mid-April, harden them off in a frame or under cloches, then set them out where they are to grow towards the end of May.

SEASON OF GROWTH. Tobacco can be grown in most soils,

but needs plenty of potash. Set the plants out in double rows 2 ft apart, staggering them 18 in. apart in the rows. Leave 4 ft between each double row. The plant should grow straight up in a single stalk ending in a flower head; so pick off any side-branches. Full growth of 5 ft or more is reached in early August: nip off the flower head. The lower leaves will then begin to ripen and change colour. Before the leaf turns yellow, break it off close to the stem. Remove the leaves higher on the stem as they too ripen.

DRYING OR CURING. When picked, thread the leaves in pairs back to back, with about a dozen leaves to each foot of string, and hang them up in a dark, airy shed for 6 to 8 weeks to dry. Take care that not only leaves but stalks are quite free of moisture – they should be stiff and leathery.

Dry leaves are brittle, and break when handled. Remoisten them by putting them in an open shed overnight to absorb moisture from the air, or lay them on the floor and cover with a damp sheet for 3 or 4 hours, after which they should be fit for handling. Now make them into 'hands' by tying together the butt ends of 30 or 40 leaves.

FERMENTATION AND MATURING. The cured tobacco can now be used, but it gives a rough smoke, and its quality can be much improved if it is further fermented in a kiln. This cannot well be done at home, and most tobacco growers send their leaves to the Scottish Amateur Tobacco Growing and Curing Association, 39 Milton Road, Kirkaldy, Fife, which will ferment members' tobacco all together in its own kiln at a small charge. The Association publishes a pamphlet on *Tobacco Growing in the British Isles* which contains useful information. After fermentation tobacco should be wrapped in polythene film and stored in a warm place for up to two years to mature.

Unfermented tobacco may be crudely prepared for smoking by taking down the dried hands when required and stirring and squeezing them in a bucket of cold water until they take on the appearance of chamois leather, then wringing out the moisture and slicing the leaves finely on a board with a blade taken from a jack-plane. The fine-cut tobacco so produced can be teased out and dried in the sun within minutes, ready for smoking, or allowed to dry slowly at room temperature.

CHAPTER FOURTEEN

BEER AND WINE

Alcohol is both a stimulant and a preservative, and all wines, beers and ciders are to some extent foods as well as digestive aids.

Wine is the drink of the southern European, ale (or cider or mead) that of the northerner; and Britons must regard themselves as northerners, even though it is possible to grow grapes in southern England. Cobbett, who in the 1800s was disgusted by the replacement of the beer tankard in the cottage by the teacup, reckoned that the labourer and his family would drink 274 gallons of ale in a year (= 6 pints a day) at a cost of about £7 5s. for taxed hops and malt; but two-thirds of this was the second brewing of low-alcohol small beer. Subsisting as he did on a coarse diet of bread, onions and potatoes helped out by bacon and cheese, beer was almost a necessity of life to the British farmworker – as was wine to the Greek or Roman slave.

A. BREWING BEER

Beer is the fermented product of three chief ingredients, barley, hops and water. It is possible, given the land, to grow and harvest a patch of barley and a row of hops, to malt the barley and to brew in one day enough beer of the best quality for a year's supply: this is best done in autumn. Or one could buy the barley and hops and malt the barley oneself; or, at greater cost, buy the prepared malt. But the dearest method of home brewing is still far cheaper than buying beer from the publican and the taste of the beer will be incomparably better. The law permits us to brew beer and make wine so long as we do not sell them.

To GROW BARLEY. Suppose you work out your annual household need for beer as 66 gallons (= 1 quart a day). This is a convenient figure, which may be reduced proportionately as you please. Now, it takes roughly 4 bushels of malt and 4 lb of hops to brew 50–60 gallons of ale in two workings, and 4 bushels of barley produces rather more than that quantity, by measure, of dry malt. To grow 4 to 5 bushels of barley you will need about ½₂ acre of good land; and on this you may grow also a single row of hops: which, however, will not give a crop until the third year. Of barley seed you will need about 1½ pecks (= 18¾ lb).

Barley needs a moderately dry and warm climate, and a light loamy soil with some lime. In the old four-course rotation it followed the turnip crop. Plumage-Archer and Spratt-Archer are favoured malting varieties. Broadcast or drill the seed in February or March; all being well it will germinate and be showing green within 3 or 4 days. If the ground is clean to start with the growth of the barley will stifle weeds and will need little attention until harvest. Allow the barley to stand until the grain is fully ripe, and cut with a scythe. Leave it in the swath for a few days, then carry it off and store it in a shed until you are ready to thresh out the grain. Tie the stalks tightly together in bundles, hang the barley up with the ears dangling and thresh on to an outspread sheet with a stick or improvised jointed flail, damaging the grain as little as possible. Store grain in a rat-proof container.

To GROW HOPS. If dried hops are easy to obtain, buy them. If not, a row can be grown on wires strained between two 8-ft end-posts. Hops need a deep, rich, heavily manured soil with good drainage, preferably in a south or south-west facing position with some shelter and a free circulation of air. Plants are obtained from cuttings or suckers taken from healthy old stools (for hops are perennials), and as they make little growth in the first year may be planted closely together in a nursing line for 12 months before planting out where they are to flower.

Your row, with a reinforced post at each end, should have

staunch poles at intervals to take the weight of the full-grown bine. Stretch a wire from post to post at a height of 6 in. from the ground, a second wire 6 ft from the ground and a third along the tops of the posts. Plant out suckers in groups of from 3 to 5, a few inches apart, at intervals of 6 ft. Female plants are needed as it is they which produce the cones, but a male or two should be included in the row. Tie a line of string vertically to the wires to provide leaders for the tendrils of the growing plants. Mulch the row and lightly fork the ground between the stools.

The cones are fit to pick when they are firm in substance and beginning to change in colour from green to light amber. Only the cones are picked, leaves and lower branches being left on the stem, which is left standing so that the nutrients it contains may find a way back into the soil.

Hops must be dried at once or they will heat and quickly spoil, especially if picked when moist. Dry by spreading the crop out on wire netting frames and letting the heat from a charcoal brazier or paraffin stove rise through them, preferably where there is a draught to distribute the heat evenly. If overheated they will lose their essential bitter oil. Dry them overnight, allow to cool (the ends of the stalks must have become shrivelled), then pack them tightly under pressure in netting sacks. Store in a dry place.

To MALT BARLEY. Soak the barley about 3 days, changing the water daily to keep the grain sweet; then 'couch' the barley or spread it out from 12 to 18 in. deep for 24–26 hours to generate the heat necessary for growth. The root will begin to grow from each grain, at which point the temperature must be lowered by 'flooring', or spreading the grain thinly over the stone or brick floor, when it must be stirred and turned two or three times a day, the temperature being kept constant at 60°F meanwhile, to inhibit the growth of shoots. In 10 or 12 days the grain will have lost its moisture, when it should be heaped more thickly to generate heat until the embryonic shoot inside (which may be seen if a grain is held up to the light) is about three-quarters the size of the grain. Now oven-dry the grains for 3 or 4 days at not more than 180°F, with warm air passing through the grains. For this they are best spread out on shelves of fine wire grid, either in an oven with an open door, or on tiers over the source of heat. The finished malt must be crisp but unburnt. The darker the malt the darker the ale that will result, so if a pale ale is wanted, the malt needs to be roasted only to a pale amber colour. After this drying the malt is screened from the rootlets and ground for use like coffee.

EQUIPMENT FOR BREWING. The essential process in brewing is the boiling of the malt in water to make the 'wort', the infusion of hops in the wort, and its fermentation by addition of yeast. How much ale can be brewed at one time depends on the size of one's boiler, to which the other utensils will conform. In the absence of the old-fashioned copper holding up to 40 gallons one could use an ex-army field kitchen of the type employed by farmers for boiling up pig-mash, or a metal tank raised on bricks with a fire beneath it. If electric power is available, a 'Burco' boiler is the thing. Lacking these, the last resort is to brew a few gallons at a time in the largest suitable vessel over the kitchen stove. Besides the boiler, one needs a mashing-tub in which to perform the first operation of steeping the malt to make the wort which is afterwards boiled, a tun-tub into which the hop-infused ale is put to 'work' or ferment, and a shallow tub for a cooler, to allow the heated wort to cool down to the right temperature for the addition of yeast. Then you must have enough casks of the right capacity to store the liquor.

Oak casks need to be fumigated with sulphur, sterilized with steam or boiling water, and then swilled round with a little of the liquor they are to contain. Barrels of inert plastic with alkathene caps and spigots are suitable as casks; a tank made of some good quality hard plastic material will serve for mashing ale, while large plastic dustbins with lids are also useful in home brewing.

BREWING. If you plan to make say 27 gallons at a brew – to include a first brewing of ale proper and a second brewing of 'small beer' from the same malt – your boiler will need to be of 20 gallons capacity, your mashing-tub of 30 gallons, your tun-tub of 15 gallons, and your coolers of proportionate size. A bushel of malt is needed for such a brewing, and $\frac{3}{4}$ lb hops. First fill the boiler with clear, preferably soft water and bring to the boil. Then put into the mashing-tub one pail of the boiling water and enough cold water to reduce the temperature to 170°F. Tip the bushel of malt and stir it well with a wooden stick. Leave for 15 minutes. Fill up the boiler again and bring it back to the boil. The 15 minutes having elapsed, add to the mashing-tub enough boiling water to make your 9 gallons of first quality ale, allowing for the soaking up of 5 gallons by the grain: say 15 gallons. Stir the malt again well. Cover the mashing-tub with clean sacks and let it stand for 2 hours. Then siphon off the wort into buckets and pour it into the tun-tub, where it will remain until the boiler is empty to receive it.

Now empty the water from the boiler into the mashing-tub to make the second brewing, of small beer. (That is, if you can bear to drink small beer, which is really meant as a thirst-quencher for the sweating labourer in the field.)

The boiler being empty, put the ale-wort from the first brewing into it, together with $\frac{3}{4}$ lb of hops well rubbed and separated. Then bring the uncovered boiler to a brisk boil for one hour or more. Put out the fire, or switch off the current, and strain the liquor into the coolers through a wire or willow basket which will retain the hops.

When this liquor has cooled to 70°F, pour it into the tun-tub. Add to this $\frac{1}{2}$ pint of working yeast (i.e. 2 oz of yeast that has previously been mixed with $\frac{1}{2}$ pint of the warm wort so that it is now active) and stir the liquor to mix the yeast well in. Cover the top of the tub. In from 6–8 hours a frothy head will rise, and will keep rising for 48 hours. After 24 hours, skim off the froth and put it into a pan; repeat in 12 hours, and then again, until the working has subsided. When the beer is cold it must be casked. The cask will be in your cellar or outhouse on its side, bung-hole uppermost, placed on a stout square frame with foot-long legs and with wedges positioned on each side of the cask to stop it rolling. The bung-hole must be slightly to one side so that excess liquid spills out in one direction, where it can be caught in a pail. Bring the beer down in buckets and pour it in through a funnel until the cask is full. The beer will resume working, and a gallon or two should be kept in hand to top up the cask. When working ceases, right the cask, bung-hole uppermost, and wedge it in place. Thrust in a handful of fresh hops. Fill the cask quite full. Put in the bung, wrapped in cloth, and hammer it home.

Meantime the small beer is made by pouring 18 gallons of boiling water from the copper on to the already used malt in the mashing-tub. Stir the grains well, cover the tub and allow it to stand for 1 hour. Siphon or draw off the liquor and put it into the tun-tub, as before. The copper now again being empty, the first ale liquor having been taken out of it to cool, tip the small beer wort into the copper, with the used hops and a ¼ lb of fresh hops, and boil briskly for an hour. Clean out the mashing-tub and strain the liquor from the copper into it. Throw away the strained-out hops and leave the small beer liquor to cool to 70°F, then pour in the working yeast as with the ale. Proceed as with the ale, but cask it while still just warm or it will not work in the barrel.

SUBSTITUTES FOR HOPS. Ale may be brewed without hops, as it was before the seventeenth century; but such ale, without the tonic, narcotic and astringent properties imparted by hops, will taste insipid. Lacking cultivated hops, wild hops may be used, or the ale may be made bitter by the use of tansy, heather tops, broom tops, wood sage, wormwood, yarrow, marjoram or buckbean. Other plants which have been used to embitter ale are sweet gale, gold withy, bog myrtle and bracken root.

TO MAKE YEAST. Brewer's yeast, or dried yeast sold for home brewing, is best, but baker's yeast can be used. Yeast can be made at home by boiling 1 oz hops in 2 quarts water for half an hour. Strain and let the liquor cool to blood heat. Add a teaspoon of salt and ¼ lb of sugar. Beat up ½ lb plain flour with some of the liquor and then mix the lot all together in a bowl. After 48 hours, add 1½ lb of boiled mashed potatoes. Leave for 24 hours, strain and bottle it, corking tightly, so that it is ready for use when wanted. The liquor must be kept near a fire while it is making, and stirred frequently. Shake before using. The yeast will keep for 8 weeks in a cool place.

BACHELOR BREWING. As little as 3 or 4 gallons of ale may be brewed at a time by using an aluminium preserving pan on a stove, a plastic dustbin, some polythene tubing and a dozen screw-top beer or cider bottles, with extract of malt either dry or bottled, and dried hops.

PALE ALE. 2 lb dried light malt extract, 3½ oz hops, 2 lb sugar, 1 pint tea, salt, yeast and water. Put hops in a muslin bag and boil for 15 minutes in a quart of water. Put the hop-water into the plastic bin with the fresh tea. Add malt, sugar, a little salt and make up to 2 gallons with boiling water. Stir, let cool to 70°F, and add yeast. Replace lid on bin and allow liquor to ferment in a warm place for about 7 days. When fermenting ceases, siphon ale into screw-top bottles containing a little sugar, to make a head when poured.

BROWN ALE. 2 lb dried dark malt extract, 1½ lb brown sugar, 2 oz hops, 1 tablespoon molasses, salt, yeast, water. Proceed as for pale ale, adding molasses at the same time as the malt.

B. MAKING CIDER

Cider (or apple wine) is the drink of the apple- or pear-growing counties as beer is of the grain-growing counties. The bitter apples grown for cider are unfit for eating raw or cooked, as are perry pears. Any apples will make cider so long as they are not too sweet, in which case a proportion of crabs may be mixed in. Late apples are better than early ones.

Gather the apples and leave for a few days to mellow, then throw out any unsound ones.

UTENSILS AND PRESSING. As fruit, unlike grain, is acid, care must be taken in the selection of utensils. Wooden casks and tubs are best, but enamel, stainless steel, silver- and tin-plate and aluminium utensils are acceptable, as are glass, earthenware and good quality plastic. To crush the apples into pulp and to press out the juice a mill of some sort and a press are required. Lacking the first you can pound one layer of apples at a time in a stout wooden tub with a baulk of timber, but a more effective crusher may be made from an old mangle with a crude box hopper superimposed, and with chromium wood-screws let into the wooden rollers in a spiral pattern with heads projecting: when the handle is turned the apples, falling from the hopper, are caught up and shredded.

The resultant 'pomace' is dumped in an open tub or tank with or without a little water, and left for a day and night so that fermentation may begin and further break down the fruit-cells.

Next, the pomace is shovelled into coarse canvas or hessian bags, which are suspended over an empty tub to drain. This drained juice is put aside for the best quality cider. The bags are then squeezed in a press to extract the rest of the juice. The more pressure the more cider, but the coarser the quality due to pips and skins.

Pressing may be done in a handscrew-operated cider press. A substitute for this can be made from two squares of thick, heavy hardwood between which the bag is pressed. One piece forms the baseplate, the other the pressure plate, and this has squared inlets cut out from two sides to allow it to slide freely up and down two strong squared wooden uprights between which it is captive, the bottoms of these uprights being fixed to the base while the tops are joined together by a cross-bar. These uprights are reinforced at the top and bottom by iron angle-brackets screwed to the outside of the frame, the bottom ones holding the frame to the base. The strengthening is essential. Grooves in the surface of the baseplate channel the juice to a point of outlet. The uprights must be high enough to allow a fair-sized bag of pulp to be placed between the two platforms, and also to permit the entry of the business end of a car jack in the space between the top platform and the strengthened cross-bar. As the jack is wound up, so compression increases between cross-bar and top plate, forcing the plate down on to the baseplate and so squashing the juice out from the bag of pulp.

FERMENTATION. The collected juice is now poured into casks and allowed to ferment spontaneously; large bung-holes will allow the cider to froth over, as it will and must do, to be caught in vessels placed beneath. To help this, the casks must be topped up continually. The longer the cider is allowed to ferment the greater its alcoholic content will be. After somewhere between 3 and 10 days it is racked off from its sediment into clean casks and stored in a cool place. Bung tightly when all fermentation has ceased. In the following spring this cider is again racked, after which it is ready for drinking. Fermentation temperature is about 50°F. Rough cider is that in which too much acetic acid has been formed through fermentation at too high a temperature. Sweet cider results when only some of the natural sugar has been converted into alcohol, suggesting too low a fermenting temperature. Champagne cider is a good mellow cider bottled (in champagne bottles) before fermentation is complete.

CIDER IN SMALL QUANTITIES. To make a gallon or two of cider, treat it as apple wine, using a wine yeast and other additives as directed for fruit wines (see next section). The washed, unpeeled apples may be pushed through a mincer or pulped in an electric juice extractor. If boiled water is added to the juice, some sugar and chopped sultanas will be needed. Apple wine may be altered in character by the addition of 1 lb of red fruit to every 4 lb of apples.

C. WINE

While the grape may seem to be the first choice of fruits for making wine, since it contains in its pips, skin and juice all the necessary ingredients for that purpose, I would advise the cottager who wishes simply to make a regular quantity of good, dry red wine for the table to concentrate upon wines made from Blackberries, Damsons or Sloes, and Elderberries, most of which can be obtained in quantity at little or no cost every autumn.

BLACKBERRIES are an easy crop to grow, either the cultivated varieties or wild brambles dug up and replanted in deep garden soil. Brambles are best grown on wires along a wall, but they make good hedges anywhere if grown along wires strained between posts. They need to be mulched, fed with liquid manure, and pruned: the old wood is cut right out after fruiting in autumn, next year's fruits being on the new shoots which must be trained along the wires to replace the old. Increase is by layering or dividing.

Country wines may, indeed, be made from many other fruits, vegetables and even flowers: with nearly all of them water is needed to increase the volume of juice and/or reduce its acidity, together with white cane or beet sugar to supplement any natural sugars which, in the fermentation process, are turned by yeasts into alcohol and waste carbon dioxide gas.

EQUIPMENT FOR MAKING WINES. As with cider, wines must not come into contact with certain metal utensils, and the same materials as for cider-making should be used.

Apart from articles found in every kitchen, the winemaker or vintager needs:

1. A large crock or jug, or plastic bin, for the first fermentation.
2. Some butter muslin or other close-woven material for

straining and for squeezing out juice from crushed pulp.
3. One or more jars or casks for the main fermentation and for storing wine.
4. One or more special glass or plastic air-locks with an equal number of holed corks for plugging the necks of those jars. (These air-locks, when partly filled with water and pushed into the holed cork in the jars, permit gas to escape from the fermenting liquor while preventing air-borne organisms from entering and infecting it.)
5. A 4-ft length of $\frac{3}{4}$-in. clear polythene tubing for siphoning the wine off the sediment or 'lees' from one container to another.
6. A glass or plastic funnel.
7. A supply of wine bottles, with new straight-sided corks.

THE MAKING OF WINE is simplicity itself. The picked fruit is washed and all unripe and imperfect fruit rejected. It is then crushed and soaked in warm water so that the natural enzymes in the fruit are enabled to attack and break down the juice-retaining pectin in the tissues. This watery pulp is then pressed through muslin into a clean crock or bin where the generally cloudy liquid is now known as 'must'. To this must are added the required quantities of water and sugar: as a rough guide, a total amount of up to 1 gallon of water and 2–3 lb sugar to each 5–6 lb of red fruit, if an efficient yeast is used. Into the must, kept at a room temperature of 65–75°F, some previously prepared active yeast is stirred – it may be baker's, brewer's, or preferably a special wine yeast. The crock or bin is covered with some thick cloth to protect the must while it undergoes its first, turbulent fermentation. Any froth is skimmed off. When fermentation steadies, the liquid is 'racked' or siphoned off into one or more large fermentation jars, the necks of which are then either plugged with cotton wool, covered with polythene film held in place by a rubber band, or preferably corked and fitted with fermentation locks. Still at room temperature, the sweetened working must is allowed to 'ferment out', that is, to work until the last of the sugar it contains has been converted to alcohol and carbon dioxide and no more bubbles rise in the jar. The wine is then racked from its lees into a clean empty jar, sealed and stored in a cool place and left for some months to clear and mature, when it is again racked off into wine-bottles, which are then corked and cellared.

So long as strict cleanliness is observed in the kitchen or still-room, only innocuous metals used, all vessels and utensils sterilized before use and all water mixed with the juice boiled and cooled, there should be no difficulty in making palatable wines in this way, especially if a good handbook is followed, such as *Home-Made Wines, Syrups and Cordials*, published by the National Federation of Women's Institutes. However, in the last 20 years home wine-making has been made more systematic by the use of substances and methods which the reader should know about, and follow where practicable.

SUGAR CONTENT OF WINE. The maximum alcohol content attainable in any wine is from 16–18 per cent. (Burgundy contains from 10–13 per cent alcohol, claret from 9–11 per cent.) The home vintager should as a rule aim at 10 to 12 per cent alcohol in a fruit wine, and since alcohol derives from the sugar present in the must, 2 parts of sugar when fully fermented yielding 1 part by volume of alcohol, it is easy to see that the amount of added sugar needed in a fermentation will depend upon the quantity of natural sugar already present in the fruit balanced against the amount of alcohol wanted in the finished wine. To discover the sugar content in a sample of the fruit juice you will need a simple instrument called a hydrometer, sold by chemists and wine-makers' suppliers with printed instructions for its use. First follow the instructions to discover the sugar content in a sample of the fruit juice poured into a measuring-glass or tall, narrow jar. Then use this formula:

Let a = alcohol desired
Let b = per cent of sugar found in juice

Let x = per cent of sugar to add to juice

Then* x = $2a - b$. Example:

If a = 12 per cent

     b = 19 per cent

then x = $(12 \times 2) - 19 = 5$ per cent

This indicates that 5 lb additional sugar will here be required for every 100 lb crushed fruit or juice, or in other words $\frac{1}{2}$ lb sugar for every 10 lb (= 1 gallon) of fruit or juice.

Sugar is best added to the must in syrup form. Having determined the amount needed, put the sugar in a pan with 2 or 3 pints water and stir while bringing slowly to the boil, skim off any scum, then cover and leave to cool. For a dry wine, just enough sugar should be added for the yeast to ferment it all out into alcohol and gas. More sugar than this would result in a sweet wine, and too much sugar would produce a wine too sickly-sweet to drink, in which case refermentation would be necessary.

STERILIZING MUST. Unwanted moulds, bacteria and wild yeasts may be prevented from breeding in the must and tainting the wine if sulphur dioxide (SO_2) is added to the must at an early stage, either in the form of potassium metabisulphite or sodium metabisulphite, $\frac{1}{4}$ teaspoon to the gallon, or in the form of Campden tablets, 2 to the gallon.

PECTIC ENZYMES (Pectinol, Pectozyme), if added to the freshly crushed fruit in the proportion of $\frac{1}{2}$ teaspoon to 10 lb fruit, will break down pectin and so reduce haziness in the wine.

ACID OR TANNIN may need to be added to some musts which are insipid or lack astringency. They may be bought

---

* The formula is easy to understand if one grasps that the amount of sugar in the juice, b, plus the amount of sugar to be added, x, gives the total amount of sugar to be converted to alcohol, a. If 1 part sugar gave 1 part alcohol, $b + x$ would = a. But since it takes 2 parts sugar to make 1 part alcohol, the formula must be adjusted accordingly, and either the total amount of sugar halved, or (which is the same thing) the per cent of alcohol desired doubled to bring the figures into relation.

in chemical form, or tannin may be extracted by infusion from oak-bark or shavings, or from tea, while citric acid is contained in lemon juice and malic acid in apple juice, both of which may be added to musts.

ADDING THE YEAST. Not only the efficient conversion of sugar to alcohol but also much of the bouquet and flavour of the wine depends on the character of the yeast used; some fruits will ferment naturally, but rather than let them do so it is better to wash them in running water and use instead of ordinary yeasts a special wine yeast, of which the all-purpose dried wine yeast is the most convenient. Yeasts will work better very often if a nutrient in the form of dibasic ammonium phosphate, urea, or malt extract is added to the must in the proportion of $\frac{1}{2}$ teaspoon to the gallon.

Instead of adding yeast direct to the must, prepare a yeast starter two or three days before making wine by diluting some juice with an equal quantity of water. Heat this to boiling, and cool. Half-fill a sterilized pint bottle with this juice and add the yeast, together with 2 teaspoons sugar, and shake. Plug the neck with cotton wool and leave the bottle at room temperature for up to 48 hours. Add this starter to the must in the proportion of $\frac{1}{4}$ pint to 1 gallon. But if as much as, say, 8 gallons of wine are to be made, prepare the starter as directed, but instead of adding it direct to the must, add it to a further $1\frac{3}{4}$ pints juice in a quart bottle and allow it to ferment vigorously before adding this larger quantity to the must. For 16 gallons, double the quantity.

TO MAKE BLACKBERRY WINE: As the berries wired along your wall become ripe, pick them daily; wash and then crush them and add to the pulping tub or crock with the addition of 3 pints boiled, warm water to every 7 or 8 lb of berries, holding back an equal quantity of water for adding later. (Optional procedures are given in square brackets.) [Add Campden tablets and Pectinol, as described.] If a large quantity of wine is to be made it may be as well to have two or more pulping tubs working alternately. Cover the tub or crock with a wad of thick, folded cloth to exclude air-borne bacteria and leave for 24 hours after the last input of crushed berries, stirring now and then. Using the legs of a chair upended on a table as a frame, improvise a strainer with a large square of folded muslin and strain the juice from the pulp through this into a bucket or bowl placed on the chair, finally squeezing the pulp hard to extract all the juice. Discard pulp, clean the pulping tub and return the juice to it, adding the rest of the boiled cooled water, to which has been added [the amount of sugar previously calculated by use of the hydrometer and the correct formula or] roughly 2 lb sugar [in the form of syrup] for each 3 pints. [Using the dried wine yeast or, better still, a special Burgundy yeast.] Add the yeast starter as described. Cover with a thick cloth and allow the fierce fermentation to proceed for 2 or 3 days. When it subsides, pour the must into casks or jars fitted with fermentation locks, filling them nearly full, and topping up with pure water from time to time as needed. When fermentation ceases, rack off each jar or cask of wine into another sterilized jar or cask, bung it tightly and store in a cool place for 6 months. After that time, again rack off the clear wine into sterilized bottles, corking tightly with corks sterilized by boiling. Store bottles on their sides in a cool shed or cellar for at least 6 months before drinking. Should the wine prior to bottling still be cloudy it may be cleared or 'fined' by the addition of a little isinglass, white sheet gelatine, or white of egg.

ELDERBERRY WINE: Gather the berries when fully ripe, strip them from the stem with a fork, rinse in running water and crush them by hand or in a juice extractor. Follow the procedure for blackberry wine, but use only enough water to equal the volume of juice extracted. Elderberries lack acid and the wine can be improved by the addition of orange, lemon or apple juice to the must.

SLOE OR DAMSON WINE: Use only ripe fruit, without stalks. Wash and crush them by hand in a tub without breaking the stones. Follow the procedure for blackberry wine, using 1 gallon water to every 8–10 lb fruit and about 2½ lb sugar to each gallon of must, but omit citric acid. Elderberries and damsons mixed in the proportions 1 part damsons, 2 parts elderberries, make a palatable wine.

Part V

Meat, Poultry
and
Dairy Produce

CHAPTER FIFTEEN

KEEPING GOATS

A. MANAGEMENT

The cow is indispensable on a smallholding when the owner has a large family, can sell surplus milk and butter and feed the skim milk to calves, pigs and poultry. Most cottagers will do better with a pair of goats, which actually produce more milk for food consumed than do cows. The common idea that goats are malign, dirty, omnivorous and smelly is quite wrong. The female goat is a clean and gentle creature, very responsive to affection. Goats are browsers, and will eat clean rough stuff that cows would not touch. Their size makes them easy to transport in the back of a car, the guard's van of a train or a trailer pulled by a bicycle. Two does or 'nannies' mated in alternate years so that each in turn is allowed to 'run on', i.e. give an uninterrupted milk supply for two years after kidding, will yield together an average of around 6 pints of milk a day throughout the year. The modern goat is normally hornless. Goats are immune to tuberculosis and seem not to get brucellosis. Their milk is medicinal and can clear up such skin complaints as eczema. Goats do, however, require incessant attention; they cannot be left untended even for a day, and must have enough land for their support – about $\frac{2}{3}$ acre each. If there is free access to rough grazing the area required for producing their fodder will be less. Goats do not have to graze at all and may be wholly stall-fed, but will still need grass or lucerne for hay or silage, some cabbages, kale and winter roots.

NEEDS OF GOATS. A Code of Practice for Goats has been drawn up by the British Goat Society, Rougham, Bury St Edmunds, Suffolk, which gives a fair idea of a goat's basic needs. The code emphasizes that a goat should be housed at night all the year round, and in bad weather. Hay must be supplied in a rack not less than 3 ft from ground level, and there must be a daily supply of fresh, clean water. Goats should not be tied up when housed, nor should horned goats be penned together. If tethered, a light chain not less than 10 ft long is needed, with at least two swivels and a light strong leather collar. The tethering peg must be moved at least twice daily. If tethered during the day the goat must be free in its shed during the night; if permanently housed it should never be kept tied up. Most of the other points made by the B.G.S. will be covered in what follows.

BUYING A GOAT. The best goat from a cottager's viewpoint is that which will give the most milk with the least trouble. What is wanted is not a show animal but a strong and healthy goat which can be expected to give about a gallon of milk a day at peak production, if not at her first lactation then at her second or third. Such an animal may be obtained from a recognized breeder, or from some isolated smallholder who happens to have one or two animals more than he can manage. Choice of breed (whether British Toggenburg, British Alpine, British Saanen or Anglo-Nubian) will depend largely upon the breeds common in the locality and the presence nearby of a suitable breeding buck or billy of that breed.

THE BREEDING CYCLE. Lactation occurs when kids are born to the female in the spring, following mating with the buck five months previously. In early autumn the doe gives signs that she is 'on heat' or 'in season' by an intermittent rapid wagging of her tail and a plaintive insistent bleating; her parts, too, will be inflamed. She should at once be taken for service to the chosen stud male. If the opportunity is missed then another will occur in 21 days' time, and another three weeks after that, and so on, until the mating season is over. If two milch goats are kept, then only one need be sent for service each year, the other if in milk being allowed to run on for another 12 months. This avoids a milkless winter followed by a glut of milk in early summer. The earlier a goat comes into milk the better, so an early mating should be provided for.

A parturient goat will kid without assistance if she is left free of restraint in a pen littered with clean straw or bracken; one, two or three kids may be produced. Examine these at once before they begin to suckle, and destroy any neuter, deformed or male kids not wanted for later eating. If all kids are unwanted males the mother's attention must be distracted while they are taken away or she will set up a continual frantic bleating. The mother will need a reviving bucket of bran mash with a spoonful of molasses, or a drink of warm ale. The slippery mucous afterbirth must be found, wrapped in newspaper and burned, otherwise the goat will eat it. Female kids should be set to the teat as soon as they can stand, for the sake of the valuable colostrum the first milk will contain. Rather than letting them run with their mother, taking all the milk they want until they are three or four months old, they should be given one good feed from the mother and then kept in a separate pen and taught to feed themselves from a bucket of milk, which they will do if first assisted with a dipped finger. There is no need for a bottle. Don't milk the doe right out for the first 3 days after kidding; and instead of cooling and using the milk taken from her, feed it back to her while still warm – it will help her over the strain of kidding and lactation.

At a few days old the kid may be given hay and allowed on to pasture, but it must have milk for 3 months, when the supply may be reduced and some calf meal substituted; at six months it should be off milk altogether. Male kids kept for meat must be castrated at 2 or 3 weeks of age. The grown kid or goatling will be ready for mating 18 months from birth, when the best available stud male should be found to improve the breed. Always breed true to type.

How to milk. The milch goat should be trained to step on to a platform raised a foot or more from the ground, where her collar can be fastened to a short tether. The milker sits on a small stool to one side, places a small milking pail on the platform, underneath the goat, and after wiping the udder and its two teats with a clean damp rag, grasps the teats firmly between the whole of the thumb and all four fingers and applies a successive and continual pressure on the teat, beginning with the first finger near the top of the teat and ending with the little finger nearer the bottom. The whole finger is used, not the tip. First one hand applies the pressure, and then the other, forcing the milk from the constantly filling teat into the 1-gallon stainless steel or tinned pail. The fore milk, the first few squirts from each teat, goes into a cup and is thrown away. The goat must be kept calm and relaxed; if she is nervy a satisfactory let-down of milk will not occur. She should be contentedly chewing the cud while being milked. If she kicks, a raised elbow will keep her hind foot from the pail. Before milking is completed the udder and teats must be stripped of the last vestige of milk. The udder is gently patted and stroked and the teats bumped in imitation of a sucking kid; this fills the teats and milking is resumed until nothing more will come. This is important, as if the goat is not fully milked out her yield will drop from day to day until she dries up. She is now ready to go back to her yard, pen or pasture until her next milking in twelve hours.

The milk must be poured through a filter into a container and cooled by placing the container under the cold tap for 20–30 minutes. The milking pail is first washed with cold water, then scalded and left upside down to drain. Goats' milk is white in colour and has a sweet, nutty flavour. Should a musty tang be noticed it indicates (a) that the milk is not fresh, or (b) the utensils used are not sterile, or (c) there is something wrong with the goat, which needs looking into. Those who have kept goats always prefer goats' milk to cows'. Its cream does not separate so readily, and it is easier to digest.

HOUSING AND MANAGEMENT. If there is no rough grazing and no meadow grass on to which the goats can be turned loose or tethered, then they must be stall fed with hay and whatever produce you can supply from your own soil. They must have a compound big enough for them to walk about in, containing a small shed for sleeping only – in which case milking will have to be done in a shelter outside the compound – or a large shed containing stalls for four or more goats, a milking parlour and fodder store. For a sleeping shed, a simple draught-free, rainproof wooden hut will do, or even a hut built up from straw bales cased in wire netting and with a floor of peat litter. A shed as big as a car garage can house four to six goats, and a 6 ft. × 8 ft. lean-to can shelter two or three. A compound must have a 4-ft high fence and a secure gate. Goats have an annoying way of squeezing through or jumping over fences and hedges and invading territory where they are not wanted, and they are inclined to strip the bark from garden trees. Every goat enclosure must be really goat-proof.

EQUIPMENT. Besides a tethering chain with ring, clip and two swivels, goats need an extra-strong leather collar, a tethering-peg, a plastic or metal drinking pail and one for feeding, a wood or metal hay-rack (indoors) and (outdoors) a rack for greenstuff, which may also be hung in wide-mesh string bags.

FEEDING. Provided the goat can browse and graze freely from May to August, all she will need for the rest of the year is a supply of good hay and enough kale and roots for the winter, with some concentrates. Instead of hay she may be fed silage. The preferred roots are cattle beet, which may be fed at any time, and mangolds, which must be kept until after Christmas when their acid sap turns to sugar. The goat will also consume all sorts of waste garden produce such as turnip and beet tops, Jerusalem artichokes and stalks, cabbage leaves, pea and bean haulms and pods, windfall apples, etc. Hay may be bought by the bale or load, and roots by the ton, from a farmer; but for obvious reasons it is best where possible to grow one's own fodder. An acre and a half of land should suffice for your two milch goats and two kids or one goatling, allowing about $\frac{3}{4}$ acre for grazing, about $\frac{1}{2}$ acre for hay and $\frac{1}{4}$ acre for roots, kale and other fodder.

THE HAY CROP. Assuming your grass to be of good quality to start with (otherwise re-seeding may be necessary), the time to cut it is while the grass is still young and sappy, and before it has reached the hard-stem, flowering stage. The younger the grass the greater its food value, and it is perfectly in order to cut grass as soon as it is six inches high and to continue cutting through the summer whenever the grass is long enough to be scythed. This short grass, cut with the scythe, makes a very light swath which begins to dry immediately in conditions where a heavy swath of long grass would dry on top but not underneath. Cutting is done in the morning, the grass raked into rows in the afternoon, then cocked and carried to the hay-store or stack as soon as it is properly dry. Such short hay packs more tightly and weighs more heavily than long hay, and needs to be fed more sparingly as it is so rich in nourishment. Two fine days together are enough to get such a crop in.

If the hay is cut long then the swath must be turned two or three times until the hay is fit, that is, until you can wring a bunch of it hard without feeling any moisture, when it can be carted and stacked. If rain threatens before hay is made, the hay should be put into cocks which shed water. Rake the hay into large heaps and build them into cone-shaped pikes with pointed tops which will let the rain slide down the outer surface and which will quickly dry out afterwards. Hay makes best on 5-ft tripods of rough poles drilled and wire-threaded at the top and spread open at the base, over which tripod is slipped, horizontally, a rough triangle formed of three poles lashed together 12 in. from their ends; this grips the tripod and holds it firm, providing a forked vent at each corner for circulation of air. The hay is piled regularly all around the frame, all loose hay being picked up from the ground at the base so as not to impede the draught of air through the central chimney. Hay so made will be ready in 1–3 weeks, but may be left out longer if need be. Rather than building a rick, store the hay under cover – an open-sided, dutch barn type shelter is ideal – conveniently near to the goats' compound. Guard against the moist hay heating up and becoming spoiled by giving any dubious hay a wider spread. A few handfuls of salt scattered throughout will absorb moisture and prevent heating.

MAKING SILAGE. Silage can be a substitute for or a supplement to hay. It is a sort of grass conserve, high in proteins, which can be made without regard to weather conditions. The fresh cut, wilted grass and other green material is tipped into a circular container and trodden down hard to exclude air, when it heats up, ferments and becomes pickled. Silage may be made in a walled clamp or a well-drained pit lined with straw; but better is an up-ended concrete drain section of, say, 4 ft across by 6 ft high, or a structure of about the same size made by bolting together sheets of corrugated iron inside old iron cart-wheel tyres. Such a silo will hold about 5 tons of silage made from lawn mowings emptied into it and trodden down hard once or twice a fortnight on a base of 8 in. of straw. (Average yield of grass per acre: 4 tons; lucerne: 20 tons.) Good silage can be made from oats and vetches, lucerne and chopped kale, as well as sugar-beet tops, the scythed verges of country lanes, etc., all of which should be wilted before use. The younger the material the less sugar it will contain; which is why molasses thinned with water (1 gallon molasses to the ton of material) is sometimes added to promote fermentation. The required heat is 100°F. To tell if the trodden silage is ready for further filling, push a metal rod into it, and if when pulled out the next day it is really hot you must speedily add more stuff; if barely warm, then leave it for another day. When there is no more material to include, seal by laying clean sacks on the top and covering them with chalk. The food value of silage of 5 to 7 lb weight is equal to 1 lb of concentrate, or 2 lb of hay, or 13 lb of roots.

CROPS FOR SOILING. On your spare $\frac{1}{4}$ acre you will grow crops for cutting green and feeding to the goats throughout the year, with an eye to the barren months of late winter and early spring. Kale and roots are the most important, but cabbage, lucerne and comfrey are all valuable.

For spring fodder grow Hungry Gap Kale. Sow the seed in

June/July where it is to crop in rows, and thin to $1\frac{1}{2}$ ft apart.

For summer fodder (to supplement grazing) sow Chicory in April/May in shallow drills 1 ft apart, thinning out to 9 in. apart. Obtain root cuttings of the perennial Giant Russian Comfrey and plant them out 2 ft apart each way in autumn or early spring. Once established the Comfrey bed will continue to produce large crops each year. Lucerne: see below.

For autumn fodder sow a row of Flatpol Cabbage in March and plant out 2 ft apart each way at end of May.

For winter fodder. Tall Green Curled Kale; Thousand-headed Kale; Marrow-stem Kale. These are sown in April/May. Perpetual Kale is a perennial, large plants result from planting out broken-off sideshoots of the parent plant. Mangolds (Mangel-wurzels) are sown in drills 18 in. apart from mid-April to early May and thinned to 9 in. apart at the beginning of June. They must be hoed often, and lifted and clamped before the frost gets at them. Cattle- or Fodder-beet require the same treatment; not as large as mangolds, they may, unlike them, be safely fed to goats before Christmas, as may Swedes.

The perennials, Giant Comfrey and Perpetual Kale should be grown on a separate plot, with Lucerne (Alfalfa) if grown. The labour-saving advantage of permanent crops is obvious. Lucerne is a deep rooter which thrives in dry conditions so long as it gets enough lime and sunlight. The plot should be deeply dug in autumn, limed and left to weather until spring, when forking and raking will give a firm seed-bed. Give the weeds time to show themselves, then kill them by hand harrowing. Sow Lucerne seeds 2 or 3 days later, not forgetting first to inoculate them with nodule-forming bacteria supplied by the seedsman, without which they may not germinate. Seeds may be drilled, or mixed with sand and broadcast. Rake in and roll. Cut once only the first year, after flowering. Three or four cuts are usual when the plants are established.

A good plan is to divide the strip of land left over from this plot into four sections (Table IV), broadcasting in each a mixture of seeds which will choke the weeds and provide

Table IV

CROPS FOR SOILING

	When sown	*Mixture*	*Season of use*	*Plot cleared and replanted*	*Season of use*
PLOT 1	*Early March*	Spring tares, maple peas and oats	May to end of June	*End of June,* plant out marrow-stem kale	Late autumn
PLOT 2	*April to June*	Hardy green turnips, giant rape, marrow-stem kale	July to September	*End of September,* plant out thousand-headed kale	Winter
PLOT 3	*June*	Broad leaf rape, hardy green turnip	Early winter	*In November,* plant out hungry-gap kale	Spring
PLOT 4	*June/July*	White mustard, hardy green turnip, Italian rye-grass, early red trifolium	Autumn	Leave stubble standing *through winter,* to sprout later	Spring

herbage for silage or soiling without further work, after raking in.

Other fodders relished by goats are: Jerusalem artichokes, carrots, boiled potatoes, swedes, whole stems of maize, kohl-rabi, sugar-beet pulp (soaked), cut tree branches, bracken, furze, crab apples (halved), elm seeds (fresh or dried), acorns (dried and halved), chestnuts and horse-chestnuts (shelled and halved), hazel nuts (unripe, with shells), raspberry canes. Avoid evergreens.

FEEDING CONCENTRATES. With dairy animals a distinction is made between maintenance and production rations, the first keeping the animal in good health and the second increasing its milk yield. Concentrates serve the second purpose, so never treat them as a main course. Some natural fodders are lacking in certain minerals necessary to the goats' health, when minerals must be added to food. My own practice has always been to keep a bin of mixed bran, crushed oats and linseed cake (say) and to mix these together with several handfuls of dry seaweed-meal. With a bowl of this mixture once a day, before evening milking, my goats have always kept healthy and milked well. Among home-grown concentrates are sunflower seeds, field peas and beans. Goats will sometimes increase their milk yield when given as a food supplement such members of the Umbelliferae family as fennel, caraway and dill. An iodized salt lick is essential.

HEALTH OF GOATS. Goats are hardy, and if well housed, well fed and kept in clean conditions, they should have few health troubles. They should be brushed down hard once a day with a stiff grooming brush, and their hoofs should be examined now and again and pared before the outer horn curves in and overgrows the soft centre pad. The knife may be replaced by or supplemented with a 'Surform' file. Dusting with derris powder will get rid of lice and external parasites. Goats can pick up the internal parasites of sheep, and should be kept off land that has been grazed by sheep within the last two years. If worm infestation is suspected, feed with carrots for a few days, and if they do not clear the worms, dose with 'Thibenzole' tablets. If a goat cuts her teat on barbed wire she must still be milked, however deep the gash; use stock iodine on the teat. Paint the whole udder with iodine in cases of udder chills and lumps in the udder. Mastitis needs veterinary attention; milk the goat frequently to prevent the udder becoming full. When hoven or bloat, keep the goat moving, massage her stomach, and drench with 4 oz of bicarbonate of soda to $\frac{1}{2}$ pint warm water. (To drench a goat, insert a small long-necked bottle in her mouth and tip it upwards at a slight angle.) Should the goat become constipated through wrong feeding or lack of exercise, give her up to $\frac{1}{2}$ pint of liquid paraffin. Diarrhoea or scouring is treated by a large dose of liquid paraffin; but it may be a symptom of more serious trouble requiring veterinary aid.

STORING MILK. Flow of milk is at its greatest in early summer, while in late winter the flow may be barely enough for daily needs. This unevenness may be levelled by storing surplus summer milk for 6 months upwards in a deep freezer in 1- or 2-pint cartons or polythene bags. Do this as soon as the new milk is cooled; if cream rises the milk may separate when thawed. Once thawed, do not re-freeze milk.

Milk may be preserved without freezing, in corked glass bottles which must be sterile and quite dry. Milk the goat into the bottles and cork them tight as soon as they are filled. Secure the corks with wire. Put the bottles on their bases in a boiler, on straw. Pack straw between the bottles and fill the boiler with cold water. Light the fire or turn on the heat, and when the water begins to boil, remove the heat and let water and bottles gradually cool. When cold, take out the bottles and pack them in sawdust in cartons. Store them in a cool outhouse or cellar. Use within a year.

(One way of sterilizing bottles is to dissolve about 1 oz caustic soda in a quart of cold water, fill the bottles to the brim and let them stand for 30 minutes, then rinse out three times with cold clean water which has first been sterilized by boiling and allowed to cool.)

B. CREAM, BUTTER AND CHEESE

Milk consists of fats, non-fat solids and water. In the making of cream, the fats are separated from the other contents by slow heating. Butter can then be made from this cream; and a superior cheese may be made from cream without the use of a curdling agent. Cheese is more usually made from the whole milk, but a less rich cheese can be made from the separated milk. Cheeses are soft and perishable or hard and keeping according to whether or not they are pressed in a certain manner. Whey, the rejected liquid from cheese, is unpalatable to humans but can be fed to pigs and poultry.

All dairying must be done in antiseptic conditions; utensils must be scalded and sterilized after use and the milker-dairyman must have clean hands and wear a clean overall in milking-shed and kitchen.

To SCALD CREAM. Goats' milk separates less readily than cows'. To make Devonshire cream, which will then make butter, pour the whole of two milkings into a shallow pan and allow to stand 24 hours in winter, 12 in summer. Then place the pan over very gentle heat, and leave it until the surface undulations thicken and a ring appears around the pan. Remove the pan to a cool place and skim off cream the next day. To make it into butter, simply beat it up by hand, adding dairy salt.

To MAKE BUTTER from unscalded cream. In this case the cream is ripened (allowed to sour and so develop lactic acid) by skimming it from standing milk and stirring each day's cream into the previous one's, leaving it in a temperature between 56 and 60°F for several days and adding no cream within 12 hours of making the butter. It should then be velvet-smooth, with a clean acid taste.

You need a simple jar churn (a 1-gallon glass churn with wooden paddles turned by a handle and gear fixed in the lid is best), a pair of wooden 'scotch hands' and a corrugated draining board, all of which must be scalded and stood in cold brine (1 lb salt to 1 gallon water) to prevent sticking. Add cold water to the cream until it resembles emulsion paint. Third- or half-fill the jar and churn it at about 53°F. When the cream begins to thicken add a cupful of cold water and churn some more. The cream will again thicken. Add more cold water and resume churning. Repeat this until the cream breaks into butter grains and buttermilk. Churn again until the grains are firm as rice-grains, then wash the butter by pouring cold water into the churn. Churn further, then strain off the liquid through a sieve. Repeat the washing until the water runs away clear. To salt the butter dissolve ½ lb dairy salt in ½ gallon water, pour it into the jar and leave for 10 minutes. Drain off the salt water and pat up the

butter with the scotch hands, pressing out all the water yet
not spoiling the butter with over-patting. Good butter keeps
over a week in summer, two weeks in winter.

To MAKE CHEESE. Cheese is a product of the separation of
the fats and solids in milk from the water (whey) effected by
rennet. Special cheese rennet (not junket rennet) must be
bought, and a special starter to speed the production of
lactic acid is also desirable. You will need some straw mats
and some moulds.

(A vegetable rennet may be extracted from the wild herb
called *Yellow* [or *Ladies'*] *Bedstraw.* And a starter may be
made at home by (1) milking ½ pint of goat's milk into a
sterilized bottle, after a first ½ pint of milk has been taken
from the goat. Seal the bottle at once and put it somewhere
in a constant temperature of 80°F, when lactic acid bacteria
will multiply and it will become sour. (2) After 3 days take a
similar ½ pint fresh milk in a sterilized bottle and pasteurize
it by standing it in cold water brought slowly to the boil and
kept simmering for 30 minutes. Cool to lukewarm and pour
into it about a teaspoon of the first soured milk. Seal and
place this new mixture in the place of the first bottle, at
80°F, and leave for 1½ days. This pasteurized soured milk is
the starter you require for making cheese. It may be replen-
ished by repeating part (2) of the process every two or three
days, adding a little of the last batch of soured milk each
time.)

A simple press can be made out of a cylindrical biscuit tin
with the lid removed and the bottom perforated, with a
slightly smaller and squatter lidded tin to fit just inside it as
a weighted follower, filled with heavy metal. Put the first tin
over a circular wood post and punch some staggered holes
in it with a ½-in. nail, then smooth the nail holes inside the
tin by hammering it upturned over an iron bar. To use, set
the first tin on a rough stand, part fill with curd, cover curd
with a piece of muslin, and place the weighted follower on
top.

SOFT COTTAGE CHEESE (1½ lb from 1 gallon) can be made

from skim milk by letting it go sour and hanging up the thickened liquid in a muslin bag to drain; but this leaves much to chance and makes at best a dull cheese. A better method is to add some lactic acid starter and a teaspoon of diluted rennet extract to a gallon of skim milk, to cut the curds when set, heat to 120°F, drain off the whey, and wash the curds in cold water.

WHOLE MILK CHEESE. For a Neufchâtel type soft cheese, put 1 gallon of sweet whole milk into a pan placed inside a larger pan containing water at 72°F. When the milk reaches 70°F, add 1 teaspoon rennet dissolved in 10 teaspoons cold water and stir it in. Allow to set overnight. Next day spread a square of clean cheesecloth over a colander and ladle the firm, smooth curd on to it. Tie the ends of the cloth together and hang the bundle up to drain. Stir the curds from time to time, and add some dairy salt. The less whey, the milder the flavour: so press or not as desired. The cheese can be eaten at once.

Try that simple cheese and then compare it with this one, made with a starter. To 1 gallon of fresh milk add 2 table-spoons starter and allow to thicken at 80°F for 20 hours. Mix in 3 or 4 pints fresh milk, then add rennet as above and stand in a cooler place for nearly 12 hours. Strain off the whey and drain the curd. When well drained the curd is ready for use. To make the same without starter a slow setting time of 5 to 10 hours is needed, and as fresh milk sets quickly it should be allowed to stand at around 80°F for 2 or 3 days before adding rennet; the milk is then kept for 12 hours at 60–70°F until set. Hang it in a bag and treat as above.

MIXED CREAM AND MILK CHEESE. Dilute rennet as described and add it to 1 gallon milk and cream at 70–80°F. Leave in a cool place to curdle, then tie curds in a cloth as before and hang up in a warm place. Open bag two or three times and turn the curds sides to middle with a knife. The next day stir the contents again and turn into the press under 7 lb weight for several hours: salt to taste.

FULL CREAM CHEESE without starter or rennet. Pasteurize the cream by heating it to 150°F, then cool to 65°F, and hang it in a bag as above. Open the bag and stir the cheese every 5 or 6 hours. When nearly set, salt and place in a mould.

TO MAKE HARD CHEESE (1 lb from 1 gallon). A simple type of hard Smallholder cheese that will keep for months when ripe can be made with 1 gallon milk heated to 85°F, with $\frac{1}{3}$ teaspoon rennet diluted in $1\frac{1}{2}$ teaspoons water. Stir rennet well into the milk for 3 minutes, cover the bowl and leave to thicken. When set, cut the curd into small cubes with a long-bladed knife. Put the bowl in a pan of hot water and heat slowly for 45 minutes to 95°F, stirring gently. Allow the curd to float in the whey for a further 20 minutes, then let it sink to the bottom for 20 minutes more. Drain off the whey and ladle curds into a square of linen scrim cheesecloth and tie into a tight bundle. Leave for 20 minutes, untie bundle and cut the curd into large squares. Tie tightly again and leave for 20 more minutes, then untie and break each piece in half. Tie tightly once again and after yet another 20 minutes break up the curd into pea-sized fragments, adding 2 teaspoons dairy salt. Line the mould with a strip of scalded scrim cloth and force the curd in hard. Apply gradually increasing pressure. After 3 hours turn out the cheese, wring out cloth in hot water and return cloth and turned cheese to mould, and press hard. The second day turn the cheese again, but put it in a dry piece of muslin and press hard. On the third day examine the rind, and if smooth and perfect, grease it with lard and bandage with calico to prevent bulging. Put on a wooden shelf and turn every day for 8 weeks to ripen. The finished cheese may be waxed by dipping quickly into 2 lb beeswax or paraffin wax which has been melted in a double saucepan.

A CHEDDAR-TYPE CHEESE may be made from pasteurized whole milk. Heat the milk to 80°F, add some starter. Leave for 30–40 minutes, then add enough diluted rennet to curdle the milk. The curd will be set in an hour, when it is cut into small cubes with a long knife and stirred gently while being heated to 100°F. The whey will separate from the curd. Allow to drain for 24 hours. Then turn the curds over and over to induce them to 'cheddar', i.e. to combine into a smooth mass. Now cut the cheese into ½-inch strips and add dairy salt. Let the salt blend with the curd, then put it under pressure for 24 hours. Store the solid cheese at 58°F on a wooden shelf for 2 weeks, turning daily. Wipe off any surface mould with a vinegar cloth. At the end of 2 weeks wax the cheese, which will keep for a year if stored at 45°F.

C. YOGHOURT

Yoghourt is a culture of the microbe *lactobacillus Bulgaricus* which when introduced into milk breeds rapidly at a temperature of 113°F. One needs a small amount of the culture for inoculation, and some means of keeping the inoculated milk at a constant temperature for 6–12 hours. Having once made yoghourt, you can continue, using some of the previous making to start off the next lot, until the bacillus ceases to be effective, when fresh culture must be bought.

Temperature may be kept at 113°F in an ordinary 2-pint Thermos flask, a gallon-sized stainless steel catering vacuum flask, or in a lidded bucket or pot placed in a haybox. Lacking these, place the sealed jug of yoghourt in a sunny window in summer or near a stove in winter.

Using a double saucepan, bring milk to 113°F, and pour into the previously warmed flask containing a spoonful of yoghourt culture; seal and shake the flask. (To make sure that milk is not contaminated by other bacteria, bring it to boiling point and simmer for 30 minutes, then cool to 113°F.) Leave overnight. If the used flask is left unwashed the yoghourt left around the sides after emptying will serve to start off the next day's heated milk. Home-made yoghourt, though just as good, will not be as thick as the bought type.

CHAPTER SIXTEEN

KEEPING PIGS

The modern cottager who decides to keep a pig should follow in the footsteps of the old-time cottagers and keep, not a breeding sow, who will be more trouble than she is worth, but a weaner, or two weaners bought at 8 weeks old. Two pigs are not much more trouble to keep than one, and they are company for each other. If one is killed and the other sold as a porker at 100 to 180 lb live weight, or as a baconer at 200 lb, the sale price of the one may go some way towards covering the feeding costs of the other. It takes 7 months to bring a weaner to readiness for slaughter for the cottager's bacon, so that one bought on 1 February will be fit for killing on 1 October, on 1 March, 1 November, and so on. The flesh will improve in quality if it is kept longer, but the high cost of feed usually makes it uneconomical to do so.

FOR AND AGAINST PIGS. Besides the high cost of their food, pigs have other drawbacks. Unless run in a well-fenced field, which they will graze, they must have a strongly built sty, which must be put up if you have not one ready-made: such a sty must be mucked out often and the dung carted to the garden. Pigs must be fed and watered twice a day. They are subject to parasites and diseases and may go sick or die. Having them killed is not exactly pleasant, and there is a good deal of work in coping with the carcass. Then too, one large hog is a lot of flesh, and may leave you with more meat on hand than you can well do with.

On the other hand, as the price of pigs varies from year to year, you may have the luck to buy your weaner cheaply. You may be able to get plenty of cheap feed of some sort. You may be able to use a pig or two to root up some rough waste ground and turn it into near cultivable condition. Your land may badly need their dung. And you may have a large, hungry family who will make quick work of two sides of bacon, two hams and all the other pork products. The pros and cons need to be weighed up carefully before you decide whether or not to keep a pig.

HOUSING. Pigs are kept either (*a*) indoors in a sty with a sleeping part and a walled yard, or (*b*) on grassland with a bare wooden or metal ark or a hut built from straw bales. Pigs keep healthier in the open, but fatten quicker indoors. A sty, whether of wood, brick or concrete must be strongly built, with no projections for the pig to gnaw and no points of leverage for his powerful snout. Walls must be smooth for ease of cleaning and floors must be of non-slip concrete insulated – as to the sleeping part – against cold and damp. This sleeping part should have an insulated roof with some ventilation but no draughts. The yard must be of concrete with a fall of 2 or 3 in. to one corner for drainage, the urine passing through a pipe set in the wall to a brick- or cement-lined sump big enough to hold a large bucket which can be emptied daily on to the compost heap. Straw or other litter is needed for house and yard, and a wood-slat floor or a low bench or bed inside the house is desirable. Pigs are clean animals, and do not like wallowing in their own muck. A non-overturnable feeding trough and a bucket for water are all that are needed, except for a block of iodized mineral hung on the wall.

Fig. 13

In the field, the pig needs an A-shaped wood or metal ark (Fig. 13) with a slatted floor if the ground is inclined to be wet, or a rough hut built from straw bales with a thatched hurdle laid on top. Straw bales are about 3 ft × 2 ft × 1½ ft in size. If the walls are made either 2 or 3 ft thick, about 30 bales will make a shack large enough for two pigs. It must be wire-netted: that is, either each bale must be wrapped separately in chicken netting or the hut must be finished inside and out with sheep netting, otherwise the pigs will pull the bales to bits. The bottom bales should stand on plastic fertilizer-bags or a base of hardcore. Instead of bales, thatched hurdles may be used, or stout corner posts fixed in the ground supporting a double skin of wire netting stuffed tightly with wheat straw or heather, the netting being wired at the top to the roof poles. Pigs should not be run on the same grass for longer than one year in case they pick up each others' parasites – which do not, however, affect sheep, geese or cattle. If the grass they eat is of good quality it will account for about 10 per cent of their food. A ring through the nose will prevent rooting, but this should not be needed if they get enough rations. Clean water must be provided in a bucket, and food in a galvanized iron trough unless cubes are fed, when they can be strewn on the grass.

FOOD AND FEEDING. The pig has only one stomach and cannot digest the coarse fodder ruminated by cattle. While pigs will eat roots and greenstuff avidly, as well as forest mast, they have been found to thrive best on meal ground from cereals. The modern standard mixed ration for pigs consists of about 50 per cent barley meal, rich in nutritive starch; from 35 to 40 per cent middlings, the offal of wheat flour which, besides starch, provides the roughage lacking in barley; and from 5 to 10 per cent protein-rich fish meal or soya bean meal, together with some minerals. The proportions of each ingredient are varied somewhat with the age of the pig. Ready-mixed meal containing all the pig needs for life and growth may be bought from the corn merchant; and the easiest way to feed a pig is to give him this meal dry ad lib, with nothing else but clean water, as farmers do. But it takes an average of $3\frac{1}{2}$ lb of balanced meal mixture to add 1 lb to the pig's live weight, and a pig bought at 8 weeks will get through as much as 6 cwt of meal before it is ready for killing. Since such meal is dear, and supplies uncertain, the cottager may be forced to cast about for cheaper foods with which to supplement a rather smaller amount of bought meal.

Cobbett, who attempted to re-introduce the maize plant into England, calling it 'Cobbett's Corn', was much in favour of fattening pigs on this corn. Unfortunately, maize cannot always be relied upon to ripen in the English climate; but maize meal can be bought and used instead of barley, up to 30 per cent of the total ration.

Taking barley as the standard against which to measure the value of other foods (apart from cereals), we find that acorns, chestnuts and peas are of the same nutritive value as barley meal, while a gallon of skim milk equals $\frac{1}{4}$ lb barley meal plus $\frac{3}{4}$ lb fish meal, and 1 gallon of whey equals $\frac{3}{4}$ lb barley meal and is high in minerals besides. Cooked potatoes are good food, 1 lb equalling $\frac{1}{4}$ lb barley meal. The following may be taken as a rough guide:

8 lb roots (swedes, mangolds, beet)
6 lb raw artichokes
12 lb greenstuff (grass mowings, cabbages, kale)
4 lb cooked potatoes
4 lb good quality kitchen swill

⎫ will replace 1 lb balanced meal ⎬⎭

Pigs will eat apples, dried stinging-nettles, dried and ground birch-bark and bracken roots; they will munch dried acorns and chestnuts whole or ground into meal, and will grow fat on the mast of beech trees. Of course, there is no need to calculate to the last ounce the value of each bucket of food one gives to a pig, but it is as well to have some idea of what nutriment he is getting and avoid obvious feeding errors.

On securing your weaner, then, and putting him in the field or in his sty, he need be fed very little for the first week while he settles down and adjusts to new food and conditions. He may then be given meal or pellets in increasing amounts until he is taking all he wants. At 8 weeks a piglet needs feeding four times a day, dropping shortly to three times then twice daily. As he grows he will eat an increasing amount, from about $2\frac{1}{2}$ lb a day at 12 weeks to about 7 lb a day at 24 weeks, this amount then keeping about the same until time for killing.

If fed wholly on meal the weaner up to about 100 lb live weight* will need not less than 10 per cent protein in the ration, this being cut down to 5 per cent after that weight, when fish meal is best replaced by bean or soya bean meal. A pig needs roughly 1 lb meal for every month of his age – the amount fed at weaning, $1\frac{1}{4}$ lb, needing to be increased by $\frac{1}{4}$ lb weekly until 6 lb a day is reached.

But that is dear feeding. If the meal is to be economized then the normal amount as above can be fed to about 50 lb live weight and then kept constant while cheaper foods like potatoes are fed as supplements, 1 lb of potatoes being added to the ration in place of each additional $\frac{1}{4}$ lb meal. After the pig gets to 100 lb, cabbages and roots can be fed on the basis that 1 lb meal = 12 lb greens or 8 lb mangolds or swedes. (But don't feed mangolds or sugar-beet before Christmas.) Whatever skim milk or whey can be fed will cut down bulk and also enable the pig to digest its present food to better purpose. Be careful not to switch too rapidly from one food to another, upsetting the animal's digestion and appetite, and not to give fresh milk one day and sour the next: better to feed all sour. A healthy pig should be adaptable as to diet, and no doubt if meal were to vanish altogether he could be fattened on potatoes, skim milk, greens, roots and maize:

---

* In the absence of scales a pig's weight may be judged by measuring in inches (*a*) right round his chest behind the shoulders, and (*b*) length from between the ears to root of tail. Multiply (*a*) by (*b*) and divide by 13, 12 or 11 according to whether the pig is lean, medium or fat. The answer gives the weight in pounds. *Example:* Length 52 in., chest 50 in. $52 \times 50 = 2,600$ in. Assuming a medium fat pig, divide by $12 = 216\cdot8$ as approximate weight in pounds.

N.B. A pig of this weight will give about 155 lb of meat, as the dead weight is about 72 per cent of the live weight.

especially if forest nuts were available in the locality for autumn foraging.

KILLING AND CURING. There are regulations covering the humane slaughter of farm animals; you may get a butcher to come and do the job, or you may have to take the pig to a slaughter-house where they will kill, bleed, gut and dismember him for a fee. Starving the animal for 24 hours beforehand is essential, but he should be allowed to drink freely.

If killed at home, the pig is secured by a noose around the upper jaw and lifted by two or three men upwards on to a strong frame such as a sawbench, where he is first stunned by a humane killer or a shot from a ·22 rifle held against the head about half an inch above the eyes. He is then immediately stuck with a sharp two-edged knife 6 or 8 in. long pushed in about an inch in front of the breastbone, in the centre line of the body at an angle of 45 degrees to a depth of about 5 in. towards the heart so that the main arteries from the heart are cut through. The heart, undamaged, will continue pumping blood for some minutes after stunning, and if the arteries are thoroughly severed blood will at first fountain from the knife wound with some force. As the pig bleeds he is hoisted up by the hind legs to assist in draining the carcass. A block and tackle rigged up at a height of 10 ft will assist in this, or the pig's hind legs may be shackled and the pig hoisted by them before sticking. Again, the pig can be killed on the ground one man standing astride him grasping the fore-feet and preventing the body from rolling with his legs while the other stuns the animal and pushes in the knife.

The carcass is either scalded in hot water at 140 to 150°F to loosen its hair and scurf, after which it is scraped smooth with a blunt knife; or it is scorched and scraped. Cobbett advises that the hair of the pig be burnt rather than scalded. 'As the hair is to be burnt off it must be dry, and care must be taken that the hog be kept on dry litter of some sort the day previous to killing. When killed he is laid upon a narrow bed of straw, not wider than his carcass, and only 2 or 3 in.

thick. He is then covered all over thinly with straw, to which according as the wind may be, the fire is put at one end. As the straw burns, it burns the hair. It requires two or three coverings and burnings, and care is taken that the skin be not in any part burnt, or parched. When the hair is all burnt off close, the hog is scraped clean, but never touched with water. The upper side being finished the hog is turned over, and the other side is treated in like manner. This work should always be done before daylight, for in the daylight you cannot so nicely discover whether the hair be sufficiently burnt off. The light of the fire is weakened by that of the day.' The straw may be helped out with dry newspapers held as torches.

The next step is to cut open the scraped carcass and remove the innards or offal. The body cavity is then washed with cold water and the carcass sawn in two through the centre of the backbone, and the head cut off. It is hung up to chill, but not to freeze, for 24–28 hours, after which it is cut up as required. Nearly all the parts of a pig are edible if prepared in the proper way, but there is a good deal of kitchen work involved in rendering the fat into lard, making brawn from the head and hocks, mincing spleen fillets and trimmings into sausages, etc. The owner of a deep-freezer will have no problem in storing pork. Others will have to hang fresh meat, lightly covered with bran or powdered borax, in a room or passage where there is a cool current of air, pickling the rest in brine until wanted for later soaking and boiling. The two flitches and hams will of course be salt-cured and dried or smoked.

CURING BACON AND HAM. For curing, 1 lb of dry salt is needed for every 10 lb of meat; and to this is now usually added $\frac{1}{4}$ lb fine powdered saltpetre for every 10 lb of salt, and some brown sugar. The first step is to soak all the pieces of meat to be cured in a cleansing brine made up with 13 lb salt mixed with 4 oz saltpetre in 5 gallons of hot water. Allow the water to cool, and put in the meat. After 1 hour take out the meat and let the brine drain off. Then lay the two flitches on a bench and into the flesh side of each rub the

curing mixture of salt, sugar and saltpetre. Treat the hams
and other pieces in the same way. Do this in a cool, stone-
floored room if possible, at about 40°F. Next make a 2-in.
thick bed of salt and press flitches and hams into it skin side
down, and lightly sprinkle with the curing mixture of salt,
sugar and saltpetre. Cover them all with 2 in. of salt, packing
it tightly around them. Leave for 5 days; take off the stale
salt; dust the flitches and hams with curing mixture again,
and again pack down in fresh salt. Do not leave for longer
than 5 weeks. Remove flitches and hams, scrape off the salt
and wipe the surfaces as dry as possible. Hang them up from
hooks in the ceiling at room temperature until fully dry.
When dry, wrap in muslin or calico against flies, and hang
up in a cool place.

SMOKING BACON AND HAMS. Bacon that is only salted and
dried is green bacon. It is perfectly good, but some people
prefer bacon to be smoked. To smoke it, instead of hanging
the flitches from the ceiling, hang them in the open chimney
over a wood fire, having first laid them on the floor and
liberally strewn and patted bran or fine sawdust upon the
flesh sides, to keep the flesh clean. Don't burn peat or coal
on the fire, nor any pine, fir or larch wood. With a constant
wood fire, the flitches will be dried in about a month; they
should be quite dry but not dried out and rigid. They must
not hang so low that the flames melt them, and they must
not be wetted by rain coming down the chimney.

In the absence of an open chimney indoors, a smoke
house can be improvised from a small outdoor shed, to
which the smoke from a brick-lined underground fire-box is
led by a drainpipe passing under one wall of the shed and up
through the floor, the flitches being hung on hooks from a
bar placed across the shed near the roof. Hams, being
smaller, can be smoked in an empty hogshead. Near the
bottom of the barrel drill two holes and through them push
a section of broomstick from which to hang the hams. Near
the top, cut a hole in the side through which to insert an iron
pan filled with sawdust and pieces of green stick. Turn the
barrel upside down, hang the hams upon the cross-stick,

push the iron pan into the opening, put a chunk of red-hot metal in the pan and cover it with oak sawdust, when the barrel will quickly fill with thick smoke. Renew the fuel from time to time and leave the ham to smoke for about 40 hours.

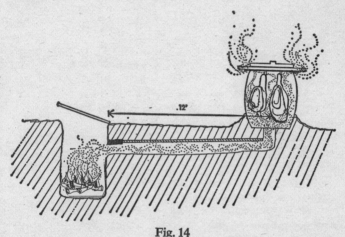

Fig. 14

Alternatively (Fig. 14), place your hogshead right side up on the ground, having first cut a hole in the bottom for the entrance of smoke. About 12 ft away from the barrel dig a fire-pit, with a trench sloping upwards from the pit to the barrel, and either covered with a 12-ft piece of wide plank-ing or lined with pipes or tiles. Fill in some earth around the base of the barrel and over the top of the board or pipes. Hang the hams from a broom handle resting in grooves in the top of the barrel, and over it place a wooden cover. Now light a fire of oak sticks and billets in the pit, and cover it with a metal plate which can be propped open for intake of air. Add fuel as needed, and smoke hams for about 40 hours.

Smoked or dried hams are best stored away in a chest or box filled with dry, sifted wood ashes where neither flies nor air can penetrate, and kept in a dry place. They will keep for more than a year in this way.

CHAPTER SEVENTEEN

KEEPING HENS

When eggs are cheap and plentiful, as they were for a time under farm-factory production methods, little or no money is saved by keeping one's own hens. But one can have the satisfaction of raising fine, healthy fowls (which those in commercial hen-batteries are not) and obtaining eggs full of nourishment and flavour, and, it may be, conserving some valuable pure strain of stock against the time when the factory system breaks down and the breeders of low-consumption, high-production hybrids go out of business. Hens will lay an egg a day each when in production, breaking off only in the autumn moulting period when they renew their feathers, and when they go broody. They eat about 4 oz of food a day, and return about 1 cwt each of valuable manure a year to the gardener. They are by no means immune from pests and diseases, but these may be kept at bay by air, sun, exercise, right feeding and clean housing.

Hens require management. You cannot buy a dozen fowls, put them behind netting in an odd corner, feed them and leave the rest to chance. They must be well housed, and have plenty of living space. A grass run will soon become a patch of mud unless it provides upwards of 25 sq. yards of grass for each bird. It is unprofitable to keep a bird beyond its second laying season. A chick hatched in February or March will begin laying, as a pullet, the following autumn; she will continue through the winter and on to the end of the next summer as a hen; but in the second winter there will be few eggs, and only one out of two hens will lay steadily through the second year. For this reason, poultry keepers who know their job get rid of half the birds before autumn and replace them with young pullets. So there must be room and housing both for the laying flock and for each year's batch of new pullets.

Pullets will come into lay without the stimulus of a cock bird, laying ordinary infertile eggs. A cockerel eats extra food, is noisy and fierce; so most people keep a few hens only, and make replacements each spring with bought day-old chicks, usually hybrids. There is much satisfaction, however, in keeping a cock and breeding one's own birds; not hybrids but pure-bred stock. The advantage of pure-bred birds is that if you have, say, one cock and six hens, they will breed true, which means that you may be sure of having in future years birds of exactly the same type as you begin with now. These birds will reproduce themselves so that there will be no need for the annual buying-in of hybrid chicks from breeders. If you choose a heavy breed such as the North Holland Blue, which I favour myself, you will have birds equally good for egg production and for meat, and hens which will with no trouble to you hatch their own brood of chicks each season from their own clutch of eggs and raise them to the stage at which they can leave her and fend for themselves. The light breeds do not go broody so readily.

If you want to breed, but prefer a mixture of two distinct stocks (it may be pure-bred Rhode Island Red cockerels mated with pure-bred Barred Plymouth Rock pullets, or vice-versa), you will have the first-year benefit of 'hybrid vigour'; but if you then continue to breed from within the flock, the features obtained in the first cross will vanish and you will end with a mongrel flock which will have lost all distinctive appearance and be valueless for sale as breeding stock. Since mongrels of any sort need as much food and attention as pure-breds, it is surely better to have the latter in the long run. Excessive in-breeding may be avoided by swopping cockerels at intervals with other owners of the breed. The activities of commercial breeders have resulted in the near extinction of the pure breeds today, so the small man who specializes in one of them may be rendering a service to future generations of mankind. The most popular pure breeds and varieties of fowl in Britain in the last decade were: Rhode Island Red, Light Sussex, Brown Leghorn,

White Leghorn, White Wyandotte, New Hampshire Red, Black Leghorn, Buff Plymouth Rock, North Holland Blue, Marans, White Sussex, Ancona. The rarer breeds, some of which may now be extinct, are: Exchequer Leghorn, Barred Plymouth Rock, Australorp, Black Minorca, Buff Orpington, Columbian Plymouth Rock, Legbar, North Holland White, Rhodebar and Welsummer.

HOUSING AND EQUIPMENT. The trend in poultry keeping throughout the last thirty years has been from free range to semi-intensive to intensive methods in which the birds are confined separately in small wire cages under artificial light – concentration camps for hens. With higher wages and the supermarket culture, working men have given up their allotments and their backyard poultry houses, so that hen-houses of the old type have gone off the market. They can sometimes be picked up second-hand, in poor condition; but one may be forced to make one's own housing. Plans for houses can be found in old books on poultry keeping and old household encyclopaedias. Pick one of these and sketch it out on paper, with whatever modifications and simplifications you desire. To build it, visit first a second-hand timber and junk yard, where doors, floors and windows can be obtained cheaply, as well as useful timber-sections and boarding. Cheap new materials are: plywood, hardboard, matchboard for walls; welded mesh for floor; clear rigid plastic, which can be cut for windows and roof-light, and linoleum and tarred felt for the roof.

Allow more space than you think you will need: if you intend to keep ten hens, make a house big enough for twenty. It is best to construct a frame of 2 in. × 2 in. timber, making each side separately and then bolting them together, cladding this frame with outdoor hardboard, asbestos board or plywood, leaving spaces for windows, a pop-hole for the birds to use and a door for human access to the inside. The floor can be of wood slats (for the dung to drop through), expanded metal or welded mesh from a builders' merchants: not wire netting. A wood or metal tray slid under the floor makes it easy to remove the droppings for dry storage.

A weather-, rat- and fox-proof house should be aimed at.
The door at the front must fit exactly, and have a lock. The
pop-hole at the left-hand end has to be kept shut at night
with a board which slides up and down within two holding
strips of batten, being hooked up by a wire or chain during
the day. Make your perches of 2 in. × 2 in. timber with top
edges slightly rounded for the hens' claws to grip, and set
them lengthwise inside the house, resting at each end in
squared sockets. The house should be draught-proof, but
with a space for air at the top angle of the roof. All wood,
inside and out, should be well creosoted on completion, and
once a year thereafter, an important sanitary measure.
Foxes will be put off by a flat metal boot-scraper with vertical bars laid in front of the pop-hole, and by loose tins containing marbles which rattle at a touch, hung from the
netting. If a nest box is included in the house, it must
be arranged so that eggs can be taken from outside the
house.

For pullets, an A-shaped ark without perches, with a
slatted floor, an opening at one end, and a roof that lifts up,
is all that is needed.

For feed, the best container is a 1-cwt size outdoor hopper
of galvanized metal; the conical top is lifted off, the bag of
meal, pellets or grain dumped into the vessel, sufficient for
several weeks, and the top replaced, when the feed will
trickle into the circular trough at the base a little at a time as
the hens peck it out. A large galvanized water-container of a
similar type, which releases water as it is drunk, is also
needed where there is no means of supplying water continuously from a pipe. The birds' need for a dry dust- or
ash-bath should also be catered for.

PLACING OF HEN-HOUSE. Poultry do well if encouraged
when chicks to eat grass; if not so encouraged, they may
never learn to graze. If there is no grassed orchard, consider
fencing off an area of garden with wire netting or – better –
chain-link fencing, setting the house in the centre of one side
and dividing the enclosure down the middle so that it is just
included in one of the two sections. Allow the hens to run

freely on one side while the other is dug and planted with a leaf crop such as lucerne or a sward of clover and mixed grasses. In the autumn, move the netting a few feet so that the house is now in the grassed section, letting the birds run on it through the next year while cultivation is done in the section which the hens have scratched, dunged and rid of pests. This alternation carries on from year to year. It permits a hay-crop to be taken from one side while the other is occupied. Or the unused side can be dug for a vegetable crop. The hens will benefit if a rough shelter is set up in a corner of each yard, with a floor of litter, to serve as a scratching shed in wet weather. Nest-boxes, which need not have individual compartments, can be set up here.

HATCHING CHICKS. A small flock of birds may be started by borrowing a broody hen from a poultry-keeping neighbour, who will be glad to lend one or two to save the cost of feeding, and giving her (the hen) either a dozen day-old chicks, or a dozen fertile eggs of the chosen breed. Don't invest in expensive brood equipment: a broody hen is far more efficient than any infra-red lamp or other gadget, however expensive. If you give her eggs to hatch out, let them be 2 oz or slightly over in weight, rather oval in shape and of strong and smooth shell texture. Test for fertility by holding each egg against a lighted lamp, rejecting any which are blemished. Allow for the loss of one or two in hatching.

Make a simple coop without a bottom to it, out of a 2-ft square box. Cover the roof with waterproof material, and make the front of vertical slats, two of which can be pulled up to make an exit for the hen, for while sitting she must be allowed to exercise herself once or twice a day, to void herself and to take nourishment. Then make a floor, over which the box is to be put, by cutting two turves of the right size, sandwiching the soil sides together, and putting them on the ground where the hen is to sit, placing the box over it. The earth between the two sods of grass must be kept damp, the moist warmth resulting from the heat of the hen's body combining with the dampness in the turf being necessary for successful hatching. Put the hen in the coop in the afternoon and let her settle comfortably before, after nightfall, gently placing the eggs under her one by one, when she will receive, shuffle and arrange them herself. As soon as the chicks emerge on the nineteenth day, pull out the brittle bits of shell from under the hen, if you wish, but don't otherwise disturb her.

CARE OF CHICKS. Chicks need frequent feeding. A first feed may be made up from chopped hard-boiled egg with some minced greenstuff sprinkled on a flat feeding-board pushed into the coop, with a saucer of milk. Repeat this two or three times, before giving them proprietary chick crumbs and water. They will soon leave the coop and peck about in the short grass on their own. Put their feed in a galvanized metal chick-trough where they cannot foul it by standing on it, and give them a pint-sized water fount which they cannot get into. Cull out any weak, sickly or deformed chicks at once. Cull again in two weeks, taking out any chicks which are not clean, bright, alert and active. Further culling must be done at 5 or 6 weeks, again at 3 months, and again if necessary when the birds reach the point of laying. There is no point in keeping third-rate fowls.

As the chicks grow they will take coarser food, with larger gaps between feeds. After a few days let them run about freely with their foster-mother, who will protect and warm them, scratch for them, and teach them how to behave. Take the mother away at 8 weeks, leaving the growing chicks undisturbed. Instead of a coop they should now have a slat-floored ark. At 12 weeks separate the cockerels from the pullets, putting them in a separate enclosure from which they can be taken for killing as needed, or sent to market at about 4 months of age.

FEEDING. In normal times ready-mixed foods in the form of mash, pellets or crumbs are obtainable from the big milling combines who deliver at weekly or fortnightly intervals. These mixes contain all the bird needs, apart from water and greenstuff: but they are dear. The cottager may prefer to buy chick crumbs or starting mash for his chicks, gradually replacing it with an all-mash growing ration or a combination of growing mash and cracked grain. At 15 weeks the birds may be fed with equal parts of mash and grain – wheat, oats, barley or maize; in which case the birds must have a constant supply of flint grit to help them grind the grain in their crops. Laying birds also need oyster shell or limestone grit to make their shells. Put these in separate, open tins.

One expert, Jim Worthington, in his *Poultry-Keeping Simplified* (1964), has shown that if laying birds are given nothing but grain and fish- or meat-meal in such a way that they can always help themselves, and provided they can get at natural herbage and earth as well as water, they will select just the quantity of each foodstuff that they need, without wastage. He points out that a cheap bulk feeder can be made by filling a clean 5- or 10-gallon oil drum with grain, putting back the 'lid' which one has previously drilled off, and knocking out with chisel and hammer a circle of small gashes all the way round the base of the drum just big enough for the grains to show through so that they can be pecked out. The drum is hung from a branch just above beak level so that the hens have to hop upwards each time they peck. Fish-meal is dear, and if hard to get, then soybean and other protein-rich meals might be tried.

If you keep only half a dozen layers, they can be fed on household scraps, potatoes and other surplus vegetables, and whey saved from cheese- or butter-making, boiled up in a pan and dried off with meal. This mash could be fed once a day, and a second feed given of grain. It is a mistake to keep more fowls than you need for egg production, in the hope of selling your surplus eggs – unless, that is, you have an arrangement with a neighbour to exchange your eggs for his bacon, milk or cheese.

GARDEN CROPS FOR POULTRY. It is not difficult to grow a patch of wheat, rye or oats in an odd quarter acre, broadcasting the seed, harrowing it in, then harvesting the grain by hand in the old way by cutting off the ripe ears with a sickle. No need to thresh; the ears being dry husk will not rot the grain. Store the heads in sacks inside metal bins and feed the birds by throwing them a few handfuls at a time, in litter, where they will have to scratch for the grain. Maize should be of the giant, farm sort. Leave the cobs on the plant until ripe, then strip them of their green fabric, hang them up in bunches under shelter to dry, then store in bins. The stems and foliage are relished by goats.

Sunflowers are valuable. After flowering, cut off a complete disc with its mass of black, oily seeds and toss it to the birds to peck at. Don't store sunflower heads, they will rot the seeds. Spread them out in the sun, face down on a wire-netting rack with newspapers underneath to catch the seeds. Store seeds in bins or jars. These seeds have a moist white kernel which can be eaten raw by humans or used in cooking.

Russian comfrey makes good summer feed, its high protein content allowing a small reduction in the amount of fish-meal fed to birds.

Cabbages and kale for poultry can be grown all the year round. Tie the whole plant to a frame or branch so that the birds have to hop up to peck at it, for they will waste leaves thrown on the ground.

Beans, carrots, parsnips, peas, swedes and turnips can all be fed to poultry; dried peas being fed as grain, dried beans being soaked and minced, carrots and turnips being sliced and hung up raw, or, like parsnips, cooked and added to mash. Poultry will eat windfall apples to excess if allowed. They will eat stinging nettles if these are first cut and wilted, the seeds being particularly liked.

A bunch of tansy, rue or mint hung in the hen-house will keep flies away.

KILLING POULTRY. The selected bird should be starved for 24 hours, to empty stomach and bowel. Hold the bird by the feet in the left hand, gathering up the wing-tips in the same hand. Take hold of the head with the right hand so that its comb lies in the palm, and extend the neck to its fullest extent, keeping up this pressure, bending the hand inwards towards the wrist, until the neck 'gives', when the bird will convulse and flap its wings (if they are not held) with decreasing violence. Let the bird hang head down, its eyes will close and wings hang loose. Hang it from a branch or beam so that the blood will drain into the space in the dislocated neck. Pluck feathers while the carcass is still warm. The method is the same for killing ducks. If your wrist is weak use the ground-bar method adopted for killing geese and turkeys. A ground-bar is a $2\frac{1}{2}$-ft long piece of wood with a notch shaved out in the centre about 1 in. deep and 6 in. long. The bar is laid on the ground with the notch over the bird's neck, comb upwards. Holding the bird's feet in the right hand, with its breast towards you, and placing your feet one on each end of the bar, you dislocate the neck with a steady pull, the sudden 'give' telling you when the neck is broken.

PRESERVING EGGS. Eggs will retain their freshness for some weeks if placed small end downwards in holed racks or shelves in a cool cellar. To preserve a glut of summer eggs for the winter months, drop them into a solution of water-glass (sodium of potassium silicate) in a large crock or galvanized pail containing a wire-mesh liner which can be pulled up when the eggs are wanted. A substitute for water-glass can be made by mixing together in a tub 1 bushel of quicklime, 2 lb of salt, and $\frac{1}{2}$ lb of cream of tartar with as much water as is needed to bring it to a consistency in which an egg will barely float. Eggs may also be preserved by coating their shells with gum arabic, afterwards packing them in dry charcoal dust. A very old method is to take eggs warm from the nest, grease them with butter or lard and pack them in boxes between layers of bran.

CHAPTER EIGHTEEN

KEEPING GEESE

The salient facts about geese are (1) that their profitable life is several times as long as that of other fowls. The same pen can be kept for ten or more years, the geese keeping up the same rate of egg and gosling production throughout. Then (2), with some meadow grass they are quite the cheapest bird to keep, for they are close grazers and will thrive upon grass, weeds and garden stuff with one meal a day of scraps and mash for egg production. One acre of meadow will support fifteen grazing geese. (3) They are even easier to house than ducks, and may be kept in a rough hut made of sods on three sides with a rough thatch roof, or with no house at all, if there is no danger from predators. Unlike ducks, they lay their eggs in one place, under a hedge or in their shelter, which should have a floor of clean litter.

Goslings may be reared by putting the eggs under hens; but a breeding pen will begin laying early in the spring, when one of the geese will soon show signs of broodiness. She will lay her clutch of eggs in a corner and sit on them for 28–30 days until they hatch, behaving afterwards just as a hen does with her chicks except that she will be an even more fearless protector of her brood. A hen can manage only four goose eggs, but a goose will sit on thirteen or fourteen. A breeding pen of three geese and a gander will produce about a hundred fertile eggs in a year.

Feed the goslings for 2 or 3 days on chopped hard-boiled egg and biscuit meal moistened with water or skim milk; then leave out the egg and get them gradually used to a simple wet mash fed two or three times a day, reducing this after 9 weeks as they become able to eat the short grass. It is widely believed that swimming water is needed if geese are to lay fertile eggs, but I have not observed this to be the case;

but they must have drinking water at all times. It is not worthwhile to insist on a pure breed: most English geese are a mixture of the Roman and Embden breeds.

One word of warning. If kept in an open situation, geese are liable to stretch their wings in a high wind and soar away. To forestall this, have the flight feathers of one wing clipped now and then with the shears.

Another warning: geese are tough birds to kill.

CHAPTER NINETEEN

KEEPING DUCKS, GUINEA FOWL AND PIGEONS

A. DUCKS

If you keep hens you will not need ducks; if ducks, hens will be superfluous – it is false economy to keep both. On balance, I think hens are preferable, but there are some situations which are ideal for ducks when, and if you like their eggs, it would be folly not to keep them. The eggs are larger than hens', with shells of a smoother grain, usually of a glaucous tinge: their whites are more glutinous. Ducks are hardy and suffer from few diseases, and a good Khaki Campbell female will lay in a year up to 300 eggs. They are easy to house, since they will do without shelter altogether; but it is best to give them a bare, perchless hut which can be shut at night – and which should be kept fastened until about 9 a.m. the next day, to give them time to lay their eggs all in one place. Unlike hens they lay always first thing in the morning, and if let out too early will drop their eggs wherever they happen to be standing, or swimming. Ducks are muckier than hens, but appear cleaner because of their water habits. Their droppings are more noisome, and their large webbed feet cause them to puddle the earth into mire whenever they are confined to a small area in wet conditions: they need each at least 12 sq. ft of room. If kept near the house, their voluble quacking can get on one's nerves and annoy neighbours. They tend to wander: I have in two separate holdings kept ducks near a stream, and they would invariably sail away after breakfast, only returning for their evening feed, having travelled quite long distances in quest of wild titbits from stream and bank.

The most popular breeds in recent times have been Khaki Campbells (for laying) and Aylesburys (for flesh); though both lay eggs and are edible. The Muscovy duck lays fewer eggs than either but needs less food and is quieter. The Indian Runner lays well, but is highly-strung. The Rouen, Orpington and Pekin are other good breeds although they may be difficult to obtain.

To RAISE DUCKLINGS. Buy a clutch of fertile eggs of the preferred breed and place them under a broody hen in a coop. The hen will treat them when hatched as her own progeny, but should be taken away when they are big enough to protect themselves and find their own food. If you want fertile eggs for further breeding, keep one drake to six or eight ducks.

FEEDING. The ducklings can be fed on chick crumbs, or breadcrumbs moistened with sour milk, or finely minced household scraps, until they are ready to eat an ordinary layer's mash ration like that fed to fowls, except that it must be damp, but not sloppy. A circular feeding trough can be made from a car tyre sliced in half. Ducks will do well on a mash of boiled potatoes and meat scraps, with bran or crushed grain. The lighter feeding breeds like the Muscovy may be simply tossed a few handfuls of bread scraps and corn. Ducks do not graze, but will eat greenstuff thrown to them, particularly the lighter-leaved plants like lettuce and comfrey. Cobbett used to feed his ducks on 'grass, corn, white cabbages and lettuces, and especially buckwheat, cut, when half ripe, and flung down in the haulm'. When he wanted to fatten them, he put them in a clean pen, where they could get at no filth, and fed them on oats, cabbages, lettuces and water.

Water for swimming is not essential for ducks, and many believe, with Cobbett, that it does them positive harm. This may be so with foul and stagnant water, but I cannot believe that clean running water does them anything but good. They must have at least enough water to wet their feet, which otherwise will become cracked and calloused.

Ducks are invaluable in clearing slugs from a garden, but do not let them in until they have breakfasted, or the quantity of worms and slugs they eat may make them ill; their long bills, too, will make short work of a row of lettuce.

Ducks are killed, plucked, drawn and cooked in much the same way as hens.

B. GUINEA FOWL

In a household where not many eggs are eaten, the guinea fowl, an African bird of the pheasant family, may be kept for its flesh, which is excellent in a 'gamey' way, and the occasional rich and delicate, hard-shelled egg. Guinea fowls are hardy, require no housing at all (they roost in the trees), and fend for themselves, requiring only some supplementary feed in the winter. The best way to secure a stock is to get a sitting of eggs and hatch them out under an ordinary hen. Keep the chicks in a run with their foster-mother for the first three weeks, feeding them as ordinary chicks, with six meals a day for the first month, dropping to four meals a day for a further fortnight, and three meals a day for the next six weeks. If left alone (killing all but one of the cock birds as required for the table) they will rear the next and subsequent generations themselves. Guinea fowl are quarrelsome, so are best not kept with other poultry. When alarmed, they make a peculiar high-pitched call, and are sometimes kept on lonely farms to give warning of the approach of strangers.

C. PIGEONS

The pigeon lofts seen at the tops of stone barns in old farm-steads were put there for breeding pairs who would feed themselves on farm leavings without trouble to their owners, who kept them for the squabs, the tender 4-week-old birds without the muscular wing development which makes older pigeons too tough to eat except in stews. As a pair of pigeons will consume more than 100 lb of grain in a year, this makes the 10 or 12 squabs they will produce rather dear, if the grain has to be bought.

But if there is a lot of free food around, it might pay to buy a pair of one of the well-fleshed breeds like the Red

Carneaux, the Giant Homer, or the White or Silver King, allowing them to fly free and housing them in a shed where the squabs are easily taken from the nest when required for pigeon pie, and where the valuable pigeon dung can be caught and collected in a droppings pit or tray littered with sawdust or peat and stored for use on compost heap and garden. Pigeons breed from February to July, laying two eggs at a time which incubate in 18 days. The breeds mentioned produce squabs weighing 14–28 oz at 25–35 days old. The standard pigeon feed is an equal mixture of maple pease, tares, durra, beans and broken maize.

CHAPTER TWENTY

RABBIT KEEPING

Rabbits are kept for their meat; but as the skins of some breeds are valuable, it is wise when choosing a breed to pick a dual-purpose animal such as the Chinchilla, the flesh of which can be eaten and the pelts dried and sold.

Consider at the outset how many carcasses you will need throughout the twelvemonth and then keep the right number of does – up to twelve – to go with the one buck you will need for breeding. Does can be bred from at 7 or 8 months, with a first mating in early February. There is a 30-day gestation period, with a litter at kindling of 5 to 8 young. The doe can be mated again in 5 or 6 weeks. At this rate she will produce say five litters a year, or thirty rabbits. Three does and a buck will produce ninety rabbits. If killed at 4 lb live weight, which produces a dead weight carcass of $2\frac{3}{4}$ lb, they will provide up to 200 lb of meat in the year, or nearly 4 lb of meat a week, together with the 90 pelts. In this time they will have got through a lot of food, since a rabbit eats daily about 2 oz for every pound of its weight. Be sure, then, that you will have enough hay, roots and greenstuff to feed this number of rabbits as well as any other livestock.

HOUSING. Rabbits will thrive in hutches under shelter, but much of the work of attending to them can be saved by running instead an outdoor colony surrounded by 5-ft high wire netting of 1-in. mesh, sunk at the bottom 18 in. or more below soil level to prevent burrowing. Each rabbit needs at least 6 sq. ft of space, with access to indoor breeding and sleeping quarters. If they can be given an area of about 10 sq. ft in a barn, or a large shed with an earth floor, fibreglass or other barriers being sunk 18 in. into the soil, then the work of clearing out their droppings daily can be saved by having a compost pit in the floor. Dig the pit 3 ft deep and fill it with compostible materials and activator; let it heat up and decompose, and then add as many earthworms as you can find. The rabbit droppings will be composted as they fall and are trodden in, making a useful manure which need be removed only once a year. Put a bale of straw in the corner for nest-making material, and cut a small door so that the rabbits can pass in and out to the feeding area.

FEEDING. Rabbits are fed mainly on mixed greenstuff and kitchen mash in summer, and on kale, roots, hay and mash in winter. Cereals are desirable but not essential: whole oats are the best. Cooked potatoes dried off with bran make a good cheap mash, and barley straw can be fed in winter instead of the dearer hay. Rabbits should have as varied a diet as possible; they should have a block of mineralized salt to lick, as well as water at all times. Suitable wild and cultivated foods are:

Cultivated

Jerusalem artichokes	Beetroot	Brussels sprouts
Cabbages	Carrots	Cauliflower
Celery	Clovers	Dandelion
Kales	Kohlrabi	Lettuce
Lucerne	Mangolds	Parsley
Pea haulm	Parsnips	Sianfoin
Spinach	Sunflower seeds	Swedes

Wild

Agrimony	Bramble	Burnet
Carrot, wild	Clovers	Coltsfoot
Common comfrey	Cow parsnip	Cleavers (goose grass)
Chickweed	Crosswort	Dandelion
Dock	Ground Elder	Groundsel
Hawk weed	Heather	Herb bennet
Hedge parsley	Ivy	Knapweed
Knot grass	Mallow	Nipplewort
Plantain	Shepherd's purse	Sow thistle
Trefoil	Vetch	Watercress
Yarrow		

The fenced outdoor area will provide much of the feed the rabbits need in summer and autumn if it is sown with a grass and clover sward – which they will nibble to a close turf – with Russian comfrey and/or giant chicory planted out at

3-ft centres. A miniature garden pond will give drinking water, and partial shade can be provided by a suitable bush or tree. The rabbits can then be left to breed naturally, but the buck must be watched and removed to a separate pen if he is troublesome. Butchering of the young rabbits can begin from 8 weeks old. See p. 170.

PELTS ARE DRIED by simply nailing them on boards left in the open air. Pack them flat in pairs, skin to skin, with a sprinkling of dried herbs or of flaked naphthalene to keep away the moth. Making them into furs is a craft that can be learned, but it is usual to send them away for curing to specialist firms.

Part VI

Wild Foods

CHAPTER TWENTY-ONE

WILD ANIMALS FOR FOOD

In giving the following account of the wild animals in Britain which may be killed and used as food it is not my intention to encourage the indiscriminate slaughter of wild animals but to give the reader information on the subject which it is assumed he will make use of only at time of real need.

GAME AND NOT GAME. There are two sorts of edible birds and mammals, those regarded as game and protected under the Game Laws, and those not so regarded. Besides the Game Laws there are Acts of Parliament in protection of certain wild creatures. Game at present includes hares, deer, rabbits, pheasants, partridges, grouse, heath or moor game, black game and (in Scotland) ptarmigan. Game is protected by the imposition of a close season (the time when the animal is breeding and rearing its young) during which it may not be hunted on penalty of fine or imprisonment. Under the 1954 and 1967 Protection of Birds Acts all wild birds, with certain exceptions, their nests and their eggs are protected. The position is now as follows.

THESE BIRDS AND ANIMALS MAY BE SHOT AT ANY TIME OF THE YEAR BY AN AUTHORIZED PERSON:

Birds

Collared Dove (Scotland only)	Merganser, Red-breasted (Scotland only)
Crow, Hooded or Carrion	Pigeon, domestic gone wild
Goosander (Scotland only)	Rock dove (Scotland only)
Gull, Greater and Lesser Black-backed	Rook
	Shag
Jackdaw	Sparrow, House
Jay	Starling
Magpie	Stockdove
	Woodpigeon

Animals

Rabbit	Squirrel	Hare

THESE BIRDS AND ANIMALS MAY BE SHOT ONLY BETWEEN THE DATES GIVEN:

Grouse, Ptarmigan (Scotland only)	12 August–10 December
Mallard, Pintail, Snipe, Teal, Widgeon	12 August–31 January
Black game	20 August–10 December (In Somerset, Devon and the New Forest, from 1 September)
Coot, Curlew (other than Stone Curlew), Godwit, Moorhen, Partridge, Plover, Green, and Plover, Grey, Redshank, Common, Whimbrel	1 September–31 January
Wild duck and wild geese in or over any area below high water-mark	1 September–20 February
Capercaillie, Pheasant, Woodcock	1 October–31 January (In Scotland from 1 September)
Deer: stags and bucks	1 August–30 April
hind and doe	1 November–1 March

By 'authorized person' is meant the landowner, tenant or person having sporting rights, or a person having written permission from one of these three, or from the appropriate public body.

To shoot game, a game licence is needed. Occupiers of land or persons formally authorized by them may shoot hares and rabbits on the land they hold. Shooting rights over land not one's own may be acquired by negotiation with the occupying farmer. Shooting is forbidden by law on Sundays and Christmas Day. For gun licences, see p. 20.

FIREARMS AND OTHER WEAPONS. A gun is not essential to the country-dweller. All guns are dangerous. But if you must have one, these are the facts about them. Airguns: Firing a small lead pellet, the modern air-rifle is a precision weapon weighing about 8 lb; fitted with telescopic sights, a ·22 model is accurate and powerful enough to be used in shooting rats, starlings, sparrows and such small deer within a range of up to 70 yards. It makes no noise, so that if one misses the target one can re-cock, load and fire again without disturbing the prey, scoring a kill perhaps with a second or third shot. Avoid an air-rifle which is cocked by a breaking action: the barrel and air-chamber are best when made in one piece. Sporting rifles: Of greater power than the air-rifle of the same calibre, lighter in weight, and better suited for killing hares, rabbits, squirrels, etc., is the miniature ·22 rifle firing the long cartridge; but its great range makes it dangerous, as even with the short rifle cartridge the bullet may carry nearly 1,000 yards, and it should not be used near any built-up area.

Shotguns: The difference between a shotgun and a rifle is that whereas the rifle has a spirally-grooved barrel which imparts a deadly velocity and accuracy to its single lead bullet, a shotgun has a smooth barrel and fires not a bullet but a spreading flight of small lead shot with a maximum killing range of about 50 yards. The most favoured is the double-barrelled 12-bore shotgun, which gives the user a second chance of bringing down his quarry. These are expensive. Single-barrel shotguns are very much cheaper. The 12-bore is good for all kinds of game. A smaller shotgun, the ·410 or 'four-ten', is used for pests and vermin, and will shoot rabbits and pigeons at 30 yards: these are cheaper still.

An old gun bought at an auction sale may have been made before black gunpowder had been replaced by modern smokeless nitro-powder. If so, it should be 'nitro-proved' by a gunsmith before it is used with modern cartridges and failing this must be used only with black-powder cartridges, for there is danger that nitro-powder ammunition will burst the gun. The owner of a black-powder gun can load his own cartridges if he has police permission to buy and store the powder. Modern cartridge cases will not take the full black-powder charge of 3 drams (in a 12-bore) with a $1\frac{1}{8}$ oz of shot. Don't then, buy a new gun that has not been nitro-proved. Look carefully at an old one.

Of other weapons only two need be considered, the longbow and the crossbow. One does not need mighty muscles to draw the modern bow, which is not the solid yew longbow but resembles rather the composite bow of antiquity which drove its arrow through the power provided by a careful blending of flexible wood, horn and sinew. The modern bow is a laminated structure built up from wafer-thin layers of very strong elastic materials such as fibreglass and specially selected hardwood, the laminations being bonded by synthetic resin adhesives while clamped over a former, the shaping being done as the bow is built up. Solid fibreglass and tubular steel bows are also made, the latter having an excellent cast. Unlike the wooden and to some extent the composite bow the steel bow is not affected by weather, does not let down during a shoot and so does not call, as does the wooden bow, for a constant raising of the point of aim.

An archer should use the strongest bow he can handle without fatigue in a full day's shoot. While the average healthy man can handle a bow with a draw weight of from 40 to 45 lb, a very strong man might do better with a bow up to 50 lb. When hunting, it is necessary to know the accurate range of one's bow, and to keep within it. The accurate range of a bow depends on its weight and the weight and design of the arrow used: the lighter the arrow the further it can fly. Assuming that bow and arrow are properly matched, and beginning with a bow of not less that 35 lb weight, allow 1 yard of distance to be shot for each pound pull of the bow, up to 60 yards – beyond which calculation errs. No archer can be sure of killing large game more than 50 yards away. Allow only 25 yards for birds, rabbits and squirrels.

More powerful and deadly than the longbow is the modern hunting crossbow. With its gun-like stock of walnut or mahogany and its laminated alloy or melamine barrel, its bow prod of 42-ton tensile alloy, strung with a Dacron bowstring, and fitted with telescopic sights, it has draw weights ranging from 45 lb in models for use by women and children to 130 lb for the hunting of really large game. 130 lb d.w. gives the arrow, or bolt, a range of 400 yards. For all-round shooting a draw weight of 100 lb is considered adequate, with a range of 360 yards. All bows are very much simpler and cheaper than guns, and of course are quite silent in action. It is not surprising that these deadly weapons should have got into the hands of commercial poaching gangs from the industrial towns who use them for large scale night raids on deer and cattle. It is illegal to kill deer with spears or arrows.

The traditional 6-ft longbow can be made by anyone with

some skill in carpentry and a vice, a clamp, a block plane, a spokeshave, a drawing-knife, a couple of files, some garnet paper and steel wool. The best wood is yew, not the quick-growing yew of the local churchyard but the mountain yew found especially in Spain and Portugal. Other woods are wych-elm, hazel and ash. With yew, the wood used is mainly that nearest the outside of the log, the light-coloured sapwood just beneath the bark. Being resistant to stretch, it makes the bow's back (the convex side when bent) while the belly is made from the darker, compression-resistant heartwood. For beginners, the best wood for a first bow, since it is easy to work, is a South American hardwood called lemonwood. The earliest longbows were made in one piece from a stave about $1\frac{1}{2}$ in. square, but the problems arising from the difficulty of getting an exactly equal bend from each limb were such as to oblige later bow-makers to begin by selecting a short wide stave which they split from end to end, bringing the two halves together butt to butt and gluing them in a fishtail joint to form one long stave with more or less balanced properties. Whatever wood is used it must be flat-grained; quarter- or off-grained staves make bows that tend to go out of line. The size of the bow is determined by the length of the arrow, and that in turn by the length of your arms. To find the right length of arrow for yourself, take a broomstick in your left hand and hold it at arm's length parallel to your left side. Standing erect and at right angles to the left arm, turn your head left and focus your eyes on the first joint of your left hand. With assistance, measure the distance from that joint to your right eye. Subtract $1\frac{1}{2}$ in. from the result. This will be your own arrow length.

To find the corresponding length for the bow, take as standard the 6-ft bow and the 28-in. arrow. For every inch or fraction of an inch subtracted from the length of the standard arrow in your own case, subtract twice that amount from the length of the bow. No arrow should exceed 28 in. Bow-strings are made from non-stretch waxed linen thread. A cable-type fishing line with a known breaking-strain may be used, when enough threads should be taken into the string to add up to 200 lb (or 150 lb for a 30-lb bow); so that if a line with a 10-lb breaking-strain is used, you will need twenty lines. Arrows may be made from straight birch dowel rods of ⁵⁄₁₆in. diameter. Instructions for making such bows and arrows are to be found in several books, including *Modern Archery* by Frank L. Bilson (1950) and *Hunting the Hard Way* by Howard Hill (1956).

For rabbit shooting, target arrows are used; for shooting grey squirrels, rooks and tree pests arrows fitted with a hardwood knob are preferred; these will stun or kill small game at close range but will bounce off if they hit the tree. Arrowheads are normally of steel, but it is remarkable that flint arrow-heads are superior to metal ones, the natural conchoidal fracture of flint giving the head about 25 per cent more penetration power. Glass, too, from broken bottles, has an extremely sharp cutting edge.

Worthy of mention as a hunting weapon is the now obsolescent throw-stick of the country labourer, carried in the coat pocket and used for knocking out small game at a distance of 30 yards. The short stick or cudgel is thickened and weighted at one end, this being the end held when the stick is thrown. It revolves in the air and the clubbed end strikes the quarry. In expert hands it is a deadly weapon, hitting its target with great force.

A. EDIBLE MAMMALS

RABBITS AND HARES. The rabbit, or coney, is not a native rodent but was brought to Britain by the Normans, who kept rabbits in enclosed warrens where they could conveniently be caught and killed for food. With the clearing of the forests, and then the rapid depletion of wild predators after the coming of the breech-loading shotgun, conditions favoured the spread of the beast, which rapidly became a serious pest to farmers and landowners. Rabbits breed mainly between January and July. They become sexually active at 4 months and have a gestation period of 28 days. In favourable conditions one doe will produce litters of six or so young at monthly intervals. Reaching maturity, the young are driven by their parents from the home warren, when they set up on their own and the breeding cycle resumes. Given maximum fertility, maximum food supply and freedom from predators, it has been calculated that one mother could give rise to 1,274,840 descendants in 4 years! Since each rabbit consumes its own weight in bark, roots, grain and greenstuff many times over each week, rabbits are formidable competitors with man for his food supply. Wild rabbits reach about 3 lb in weight, tame ones double that figure.

HARES belong to the same genus as rabbits, but they are much larger; instead of grey their colour is reddish-brown; they are longer-legged and have black-tipped ears. They do not burrow nor are they gregarious like rabbits, but produce their young – with eyes open and complete with fur – above ground. They lie singly in their form in the fields, and when started, career in huge leaps, propelled by their kangaroo-like back legs. Their speed protects them from enemies. Both rabbits and hares are esculent, the flesh of the rabbit being white and tender, that of the hare coarser, redder and more 'gamey'. Hare is messier to prepare for the pot.

TRAPS. Shooting apart, both rabbits and hares may be taken in traps, of which the most used is the looped-wire snare. Ready-made snares may be bought for a few pence in country towns, but they can be as well made at home. One trap can be made from 6 ft of brass wire of the correct gauge, folded into four with its strands twisted into a single cable. Do this by fixing one end of a galvanized steel 'S' hook to a heavyish weight and pushing the other through the two loops formed by the doubling of the strands of wire. Hold one end of the wire, let the weight hang down and set it spinning rapidly, having first set a small brass eyelet in the double loop. Now thread the upper end of the cable through the eyelet. Take the upper-end, bend back about an inch of it and twist it round the strands to form a $\frac{1}{4}$-in. loop. Next take a 3-ft length of strong string, double it and put the doubled end through the small end loop on the wire. Slip the two loose ends through the looped end, like a slip-knot, and pull tight. Now knot the free ends of the string round a wooden peg about 1 ft long, notched at the top to receive the string and pointed at the other end for driving into the ground. This completes the snare. Only one more thing is to be done: if the loop runs all the way it will tighten cruelly round the neck of the rabbit which runs its head into it and slowly strangle it. To prevent this, knot the wire 6 in. from the brass eyelet so that it will not pass through it: the noose cannot then be tightened beyond that point. The snared rabbit will either break its neck or will wait quietly in the snare until released and dispatched. Before using a new snare, hang it up to weather for a day or two.

To SET A SNARE, find the rabbit's habitual run. Choose a point half-way between two of its sets of footprints (rabbits are leapers) and knock the peg into the ground about 1 ft to one side of the run. Rub your hands on the ground to get rid of the human smell, to which rabbits are sensitive, and open the noose to about 4 in. across. Find a pointed stick about 6 in. long, sharpen one end to a point and make a groove in the other end to support the noose. Press the wire near the noose into the groove in the stick, and push the pointed end into the ground. The noose should now stand upright across the run, its lower end being a hand's breadth high for rabbits, for hares half a hand higher. Visit the snare first thing in the morning and last thing at night.

In *The Gamekeeper at Home* (1878), Richard Jeffries stresses that practice and skill are essential to the successful snaring of hares. The loop of the snare must be neither too large nor too small, it must be set at the right height above the ground, which is 'measured by placing the clenched fist on the earth and then putting the extended thumb of the other open hand upon it, stretching it out as in the action of spanning, when the tip of the little finger gives the right height for the lower bend of the loop – that is, as a rule . . . A hare carries his head much higher than might be thought; and he is very strong, so that the plug which holds the wire must be driven in firmly to withstand his first convulsive struggle. The small upright stick whose cleft suspends the wire across the "run" must not be put too near the hare's path, or he will see it, and it must be tolerably stiff, or his head will push the wire aside.' However, some people believe that it is wrong to drive the holding peg too firmly into the ground, because a hare caught in anything firm will scream, whereas if it is caught in something that moves it will go on jumping about in an effort to get free, and so will quickly tire itself out and lie still. But clearly the peg must not be so loose that it is pulled out of the ground and the hare allowed to run away with it.

Jefferies further describes how hares may be caught with no aid but the hand, especially in winter weather when the cold makes all wild animals 'dummel', or slow to move. 'It requires a practised eye, that knows precisely where to look among the grass, to detect him hidden in the bunch under the dead, dry bennets. An inexperienced person chancing to see a hare sitting like this would naturally stop short in walking to get a better view; whereupon the animal, feeling that he was observed, would instantly make a rush. You must persuade the hare that he is unseen; and so long as he notices no start or sign of recognition – his eye is on you from first entering the field – he will remain still, believing that you will pass.

'The poacher, having marked his game, looks steadily in front of him, never turning his head, but insensibly changes his course and quietly approaches him sidelong. Then, in the moment of passing, he falls quick as lightning on his knee, and seizes the hare just behind the poll. It is the only place where the sudden grasp would hold him in his convulsive terror ... and almost ere he can shriek (as he will do) the left hand has tightened round the hind legs. Stretching him to his full length across the knee, the right thumb, with a peculiar twist, dislocates his neck, and he is dead in an instant ... It is very easy to sprain the thumb while learning the trick.

'A poacher will sometimes place his hat gently on the ground, when first catching sight of a sitting hare, and then stealthily approach on the opposite side. The hare watches the hat, while the real enemy comes up unawares, or, if both are seen, he is in doubt which way to dash.'

In the absence of a snare, make a deadfall trap by propping up a stone slab with a baited twig, to reach which the victim must reach up and, moving the twig, cause the slab to fall and crush it. Another trap is the figure-four trap, in which the three wooden supports used form a figure 4. First, a sharpened base support is stuck vertically into the ground; then a cross-piece is notched in the centre to fit a smaller notch in the centre of the upright base support; next the lower end of a slanting lateral spar is fitted into a notch in the end of the cross-piece, its centre resting on top of the base-piece. This frame is kept in position by the weight of a stone slab leaning against the lateral spar. A bait is placed on the cross-piece, and when this is snatched the edifice collapses on the animal.

KILLING RABBITS AND HARES. Rabbits are dispatched either by a sharp blow behind the ears or by dislocating the neck by pulling. In the latter case, hold the rabbit's hind legs in the right hand, letting it hang head downwards, then grasp the head with the left hand, fingers under the chin and the thumb extended behind the ears. Stretch the rabbit firmly but without force and then press the chin up and the base of the head down. The force needed varies with the size and age of the animal, it must be enough to dislocate the neck at once without pulling away the head.

To kill by a blow, hold the rabbit by the hind legs when, in hanging down, it will normally raise its head which makes it easy to kill with a karate chop at the back of the neck.

PAUNCHING. When killed, the rabbit or hare should have its hind legs slit just above each hock between the sinew and the bone so that a skewer or cord can be passed through, enabling it to be hung up for the blood to drain into the neck. Before it can be skinned it must be paunched. Having covered the kitchen table with newspaper, lay the animal on its back with its head away from you and slit open the belly with scissors or a sharp knife. First cut through the pelt, laying it carefully aside, then make a shallow cut in the inner skin. Take out the intestines, wrap in newspaper and discard, saving only the heart, liver and kidneys. Wipe with a damp clean cloth. After paunching, skin and cook a rabbit as soon as possible, but hang a hare from 3 to 6 days according to the weather.

TO SKIN A RABBIT, first cut the skin round the lower joint of the hind legs, completely severing the paws. Push the hind legs up to the belly, holding tight to the pelt, which will pull away without trouble. Ease the pelt over the rump, cutting off the tail if necessary. Sever the front paws at the lower joint and pull the front legs through the pelt. Cut off the head with the pelt complete. A rabbit is trussed for cooking by laying it on its back, bringing the back legs inwards to the middle of the belly and fastening them with string, the front legs being treated in the same way.

SQUIRRELS. The native red squirrel is a harmless animal of vegetarian habit. It is disappearing, being displaced by the grey squirrel which was let loose at the end of the last century, and which eats not only woodland foods but cultivated nuts and garden fruits and the eggs and young of birds. It lives in nests or 'dreys' in trees, sometimes in tree-holes and sometimes in old nests of magpies and crows. It has two broods each year, each with about five young. Its flesh is edible and in America is considered a delicacy, as it resembles rabbit in flavour. Paunch, skin and truss as for the rabbit, and use within 3 days. Cook as rabbit.

VENISON is the culinary term for the flesh of deer of all kinds, and is sometimes extended to include goat-flesh, which is similar. The flesh of the buck is in season from June to the end of September, that of the doe from October to January. The flesh of the forest-fed animal is considered the choicest, but since venison improves with age an old animal is preferred to a young one. Skin as soon as shot. The meat is improved by moderate hanging – not more than 2 weeks in a cool, dry place. Only the haunch and saddle are worth roasting, the rest being stewed or made into pies. When hung the meat must be examined each day for signs of tainting by running a small knife into it; if when withdrawn the blade has an unpleasant smell, the tainted parts must be washed with warm milk and water, dried, and covered with ground ginger and pepper – which must be removed before cooking. The haunch is cooked by rubbing it all over with fat and covering it with greased paper over which is then brushed a paste of flour and water, this being covered with more greased paper tied in place with string, the whole then being roasted before a glowing fire or in a moderate oven for 3 to 4 hours, with frequent basting. Half an hour before serving, the paper and paste are removed, the joint dredged lightly with flour and basted with hot butter until it has a good brown colour.

HEDGEHOG. These are best kept as pets in the garden, where they will drink milk from a saucer and clear the ground of slugs and snails, or, in the house, will rapidly clear a plague of cockroaches. They hibernate in winter in hollows scooped in the ground and lined with leaves. Baked hedgehog is a renowned gipsy dish. The paunched animal is cooked whole, first being rolled into a ball and covered with clay: this is baked in the embers of a wood fire and then cracked open, the spines and skin coming away with the clay shell, leaving the succulent flesh exposed. Serve with roasted whole potatoes.

EDIBLE DORMOUSE. This animal was introduced to England about 60 years ago. Prevalent in and around Hertfordshire, it is as big as a rat, with a bushy tail; in colour it is like the grey squirrel, with brown on the back and brown rings round the eyes. It lives in holes in trees or buildings. Nocturnal in habit, it can be caught in a trap baited with apple. Treat as rabbit.

B. EDIBLE BIRDS

The usual way of obtaining birds for the table is to shoot them. Birds may also be taken by netting, by means of cages, by means of bird-lime, by means of inebriants and by the employment of hawks or falcons. The last, being a recondite form of sport, need not concern us.

The netting of birds is usually carried out at night, all that is needed being some lengths of fish netting draped between trees or on poles, and an electric torch. The light is shone on the net and the awakened birds dash towards it and become entangled in the mesh, from which they are removed and killed. This way of taking birds is illegal. Cage traps for birds are made of wire netting on wood frames and work on the principle that access is made easier than egress. Such a trap, baited with grain, is left in some place frequented by the kind of bird one is after. It must have room enough for a good number of birds and a small door through which trapped birds can be withdrawn. Keepers use such traps to obtain game birds for breeding at the end of the season. Liming is an ancient way of catching birds by spreading bird-lime on the ground, on the twigs of trees or on wire netting, but is now illegal. The use of inebriants, also now illegal, involves soaking a quantity of wheat, rye, barley or peas in a narcotic liquor, strewing them on the land where the birds are to be taken and picking up the befuddled or insensible birds before their systems have recovered from the intoxication. A game-book of 1686 recommends soaking the grain in an infusion of nux vomica, a seed containing strychnine; in the lees of wine; or in the juice of hemlock mixed with the seeds of henbane and poppy. Poachers have long known this trick, but usually soak the grain in rum or whisky. The dregs of the must of a home-made wheat wine have been found equally effective. A way of trapping pheasants in winter is to ram a large bottle neck downwards in hard snow and withdraw it, leaving a firm impression of the bottle. If the narrow part of the impression is filled with dried peas soaked in water a pheasant, in taking the peas, will lean further and further over until it topples with its head in the hole.

Game birds are killed by pressing the crossed thumbs down on the back of the skull. This is less easy the older and larger the bird. The old fowler's method is to bite the skull between the teeth. Alternatively, hold the bird by the body and sharply rap the back of the skull and neck over the toe or heel of your boot.

PREPARING POULTRY AND GAME. These instructions apply to domestic poultry as well as to game. To pluck a bird (while the bird is still warm; if it has gone cold, dip it into hot water for 1 minute) suspend it, if heavy, from a hook, or hold it in the left hand if light, and pull out the wing feathers first, then continue across the back and breast. To singe, wrap the bird in a sheet of dry newspaper, place on a metal grid and ignite the paper. Or, hold the bird's neck in one hand and with the other apply the flame from a twist of lighted paper. To draw, cut off the head and neck, remove windpipe and crop, cutting the skin attaching the crop to the carcass. Sever legs 2 in. below joint, bend and break them off, then pull out the tendons one at a time, having wound the ends round a skewer. Cut the skin between vent and tail and with two fingers inserted loosen the bird's innards from the stomach walls, being careful not to break the gall-bladder. Gently pull out the whole innards. Wash the bird under running cold water and dry with a cloth. Wash liver and heart in cold water. To disjoint, cut the skin around the thigh, then cut through the joint and remove leg and thigh. Cut through base of wing and remove by cutting it away from the breast at the joint. Remove the other leg and wing and split the bird in two with a knife, separating breast from back and ribs. To truss, place bird on its breast and cross the ends of the wings over the back of its neck. Hold the legs, and pressing the thighs into the sides of the bird, pass a needle and thread through the bottom of one thigh, through the body and out through the other thigh. Turn bird on its breast and pass needle through the middle of pinion, fastening with it the skin to the back and bringing the string out through the other pinion. Pull ends of string tight and tie together at the side. Turn bird on its back, pass the needle and thread through the skin, over the bottom of the breast-bone, over one leg and through back of bird, and tie it over the legs at the other side.

Some birds are too small to pluck, in which case they are skinned by slitting the skin down the back and gently working it away from the bird, complete with feathers. To make this easier, the wings, and the feet from the knee-joint, are first removed.

BLACKBIRD (In season from August to end of January). Skin and draw the birds and stuff them with forcemeat. Line a pie-dish with thin slices of steak, put in the birds cut in half, season with salt and pepper. Half fill the dish with stock, cover with pastry and bake in a moderate oven.

GULL. Gulls need to be caught alive and fed for three weeks on barley and buttermilk to take away their fishy taint. Pluck, singe and draw and wash the bird well. Stuff with thyme, cover breast with rasher of bacon, wrap in greased paper and roast in moderate oven for 40 minutes.

LARK (November to February). Pluck and singe, cut off feet and remove gizzards. Truss birds by holding them in shape and skewering them six at a thrust. Brush them over with hot fat and roast before a hot fire for 10 minutes, basting the while. Serve on toast. To make a lark pie, pluck; singe and draw the birds and stuff them with mushrooms and bacon chopped small, parsley and sweet herbs, pepper, salt, nutmeg and mace. Place in a pie-dish lined with bacon, season, cover with bacon rashers, and then with pie-crust and bake. When cooked, take off crust, remove the bacon, pour in some gravy or stock, warm and serve.

MOORFOWL. Skin rather than pluck. For roasting, truss them with heads under their wings and stuff with breadcrumbs and butter. Baste frequently. Cook for 30–35 minutes.

PARTRIDGE. Hang for not more than 4 days. Stuff with vine leaves sprinkled with brandy; roast as pheasant; remove and discard vine leaves. Old birds should be casseroled.

PEACOCK. The hen is preferred. Truss like chicken, but retain the unplucked head. Cook as pheasant.

PHEASANT (Early October to February). Hang for up to 3 weeks. The hen is preferred. To roast: pluck, clean and draw, keeping tail feathers for garnish. Put 2 oz butter and a shallot inside the bird, wrap two rashers bacon round the breast. Cover with greaseproof paper and roast for up to 1 hour, according to size. Dredge with flour, baste, and return to the oven to brown. Garnish with watercress and serve with bread sauce.

PIGEON (March to October). Draw the pigeon, which should be a young one, as soon as killed and cook within 4 days. Pluck, singe and clean, wash well and dry. Cut off the head close to body, saving the neck to make gravy, and sever feet at first joint. With a needle run a fine string through both legs and body, bring the string up round the rump and tie the legs up to it. Boil, braise, fry, grill, stew or put into a pie.

ROOK. The birds are skinned rather than plucked. Cut out backbones and put aside. Avoid breaking the galls. Season with pepper and salt and place birds in a deep pie-dish with $\frac{1}{2}$ pint of water and a piece of butter. Cover with a light crust over which place a sheet of buttered paper, and bake for $2\frac{1}{2}$ to 3 hours.

SNIPE (November to February). A small game bird, the weight varying from 2 to 10 oz. Roast for a few minutes on a spit over charcoal, without cleaning the bird. Or place on a slice of buttered toast in a roasting tin and roast for 15–20 minutes. Serve with lemon and melted butter on toast.

SPARROW AND STARLING. Treat as Blackbird.

SWAN. The young cygnet is eaten in September for preference; it will weigh up to 18 lb. Pluck and singe, draw and truss in the usual way, and roast.

WILD DUCK. Truss for roasting. Place in a deep baking-tin with $\frac{1}{2}$ in. of boiling salted water in the bottom and bake for 10 minutes to remove fishy taint. Dry the bird, dredge with flour, baste with butter and roast in front of fire for 20 minutes, or bake for 20 minutes in a moderate oven, basting frequently.

WOODCOCK (September). The male bird is preferred. Hang for 2–8 days until mature. Pluck the body, skin the head and neck. Do not draw. To roast: truss, brush with melted butter and cook as snipe.

WOODPIGEON. Hang for a few days. Cook as pigeon.

C. EGGS, FROGS AND SNAILS

The eggs of most birds are edible, but wild birds' eggs are often too small and too bitter in taste to be worth gathering, while most people today feel some compunction about taking eggs, which is in any case now illegal, with a few exceptions. Not long ago there was a huge trade in plovers' (lapwings') eggs, which fetched high prices on the London market; it is now forbidden by law to buy or sell them. The eggs of five species of gull may be sold for human consumption.

That snails and frogs are edible is well known. Henry Mayhew in *London Labour and the London Poor* records a conversation he had *circa* 1850 with a young street seller of birds' nests and wild foods who sold snails and frogs to Frenchmen. The Frenchmen, it appears, always picked out first the yellow-bellied frogs – 'the others they're afraid is toads'. Of snails:

'The snails I sell by the pailful – at 2*s*. 6*d*. the pail. There is some hundreds in a pail. The wet weather is the best times for catching 'em; the French people eats 'em. They boils 'em first to get 'em out of the shell and get rid of the green froth; then they boils them again, and after that in vinegar. They eats 'em hot, but some of the foreigners like 'em cold. They say they're better, if possible, than whelks.'

The two kinds of frog were clearly the Common Brown Frog which abounds in Britain, and the Green or Edible Frog. Frogs are caught in ponds and slow streams by means of nets and a wooden rake. It is the hind legs which are eaten. They are in season in Lent. Cooking is simple, the frogs' legs being floured, seasoned and lightly fried in a little butter for 10–12 minutes, or they are dipped in melted, seasoned butter and grilled for 5 minutes.

Of snails (Fr. *escargot*), those eaten are the Edible or Roman Snail, a large tawny species, and the Common Garden Snail. If collected wild they should be starved for 2 days, or until they stop foaming, before using. Cook by throwing them into a pot of boiling water and boil for 20 minutes. Remove from shells, take out the small intestine, then cook in a saucepan with butter, flour, stock and various seasonings.

D. FRESHWATER FISHES

Britain has more than 300 rivers, all of them fish-bearing except where the pollutions of industry have killed off the fish. Besides the rivers are innumerable streams and brooks, canals, reservoirs, gravel and clay pits, all of which may contain edible fish, sometimes of large size. Of river fish, the salmon and the trout, which belong to the same family, are the most prized both for sport and for eating. They are jealously preserved, and are caught with a fly mainly by skilled anglers. All other fish are known as coarse fish, and are caught with submerged baits by unprivileged but no less devoted anglers, who do not as a rule sell or eat their catch but weigh and measure it and throw it back into the water. While not such delicate eating as fish of the salmon family, coarse fish are nevertheless edible. Some of them have an insipid or 'muddy' flavour; the latter may be removed by keeping the fish alive in a tank of clear water for a few days before it is eaten. Most fish taste well when newly caught if gutted, wiped and lightly grilled over charcoal, with a little butter and some herbs. Some fish attain a remarkable size if left in peace to grow, for the life-span of some fishes, like the pike and the carp, may exceed that of a man. Pike have been caught weighing over 47 lb, carp over 44 lb, and barbel more than 14 lb. A trout long on the records list was 35 lb, while salmon will grow up to 40 or 50 lb.

Nor is there a necessary relation between the size of a stream and the size of the fish to be found in it, for since the smallest stream connects with a larger one, and that eventually to the broad river which runs to the sea, a large fish may find its way upstream to a small brook. A stream near your homestead small enough for you to jump across may contain a variety of fish, especially if it has some deep pools and clean, aerated water. And among the fish may be the occasional one of large size.

As with game, there is a close season for river fish: for coarse fish 15 March to 15 June; for trout (rod) 1 October to 1 March; for salmon (rod) 1 November to 31 January and salmon (net) 1 September to 31 January (but in Scotland, from 1 October to 14 March). As local variations occur, some of these dates are approximate. In some places where the preservation of salmon and trout is given priority, there is no coarse fish close season. Further, there are legal enactments prohibiting certains ways of taking fish.

There is a broad distinction between those fish which live in deep, slow-flowing water with weedy stretches, deep pools, bends and canal-like sections (bream, carp, perch, rudd, tench, pike, ruffe and roach) and those which prefer fast shallows and middling deep reaches of medium flow (dace, chub, bleak, barbel, gudgeon). The following rough chart (Table V) names and describes the principal freshwater fish and notes their degrees of edibility.

The baits used by anglers are varied, ranging from items in or near to the fishes' natural diet at the one end to macaroni, sausage and cheese at the other. Some anglers employ 'dopes' and flavourings like oil of fennel and aniseed oil to attract fish. The standard groundbait for getting fish to feed is ground-up stale bread. But there are other ways of taking fish than by angling, most of them now illegal. For example:

FISH TRAP. A willow or wicker framework covered with net, and with a retracted mouth which gradually narrows in the manner of a lobster pot, making it easier for a fish to enter than get out, is placed at a strategic point in the stream and inspected from time to time for captive fish. Portable fish traps of wire mesh, 3 ft and 4 ft long, of a barrel shape, with funnels from each end, are still made by S. Young & Sons Ltd, of Misterton. They are used from April to October or November, baited with worms, offal or herring-guts

Table V

FRESHWATER FISH

Fish	Description	Habitat	Food	Bait	Size	Edibility
Barbel	Green-brown back, brown-yellow sides. Long body, four barbules at mouth.	Fast rivers, sandy bottom.	Worms, insects, plants.	Lobworm, maggots, cheese.	12–24 in. 6–8 lb	Poor
Bleak	Elongated body with bluish-green upper sides, under-surface silver, projecting chin.	Slow-flowing streams	Frogs, newts, spawn.	Maggots, paste.	6–7 in.	Fair
Bream	Bronze with grey-black back. Deep body, compressed behind ventral fin.	Lakes, ponds, slow rivers.	Plants. Bottom-feeder.	Worms, maggots.	12–24 in. 5 lb	Fair
Carp	Long and sleek with large scales. Barbule at mouth-corners. Grey-black back, yellow-brown sides.	Shallow weedy lakes, quiet reaches of slow rivers.	Plants, worms.	Bread, lobworms, potato.	12–24 in. 2–3 lb	Good
Chub	Brassy, large tail; cheeks and gill-covers yellow.	Rapid waters, with clear bottoms. Lakes.	All kinds.	All baits.	10–18 in. 3–5 lb	Fair
Dace	Elongated body similar to Chub, but more silvery in colour.	Swims in shoals in deep, slow waters.	Surface-feeder.	Maggots, worms, bread.	8 in.	Poor

Fish	Description	Habitat	Food	Bait	Size	Edibility
Eel	Snake-like body fringed with fins.	Crevices in river bottom.	Frogs, newts, spawn, fry.	Small dead fish, worms.	18 in. 5 lb	Good
Grayling	Silver-grey. Resembles Trout, but has long dorsal fin.	Fast-flowing reaches of clean rivers.	Insects, worms.	Maggots, worms, bread.	4–5 lb	Esteemed
Gudgeon	Looks like a small barbel, with one pair barbules. Olive-brown upper parts spotted with black.	Running water with gravel bottom.	Worms, molluscs, insects.	Maggots, worms, pastes.	3–6 in.	Esteemed
Perch	Deep body, spiny dorsal fin, barred markings.	Still waters, ponds.	Insects, worms, small fish.	Worms, live bait.	2–4 lb	Esteemed
Pike (Jack)	Oblong, olive-grey body, large jaws.	Deep, slow-flowing waters.	Frogs, small fishes, animals, birds.	Live and dead bait.	10–20 lb	Good
Roach	Dull green back, silvery body.	Medium-flow rivers.	Shrimps, larvae, crustacea.	Maggots, bread, paste.	10–12 in.	Poor
Rudd	Deep coppery body, lower fins scarlet. Protruding lower lip.	Sluggish water, rush and weed.	As roach.	Maggots, grubs and worms.	6–12 in. 1 lb	Passable

Table V—continued

Fish	Description	Habitat	Food	Bait	Size	Edibility
Ruffe (Pope)	Olive-green mottled with brown.	Slow rivers.	—	—	5–6 in.	Esteemed
Salmon	Blue-green-grey back, silver-white underneath, black-spotted.	Fast running rocky and sandy streams.	Minnows, trout-fry, worms, flies.	Worm, fly.	10–20 lb	Highly esteemed
Tench	Thick yellow-brown body, barbels, rounded fins and tail.	Deep, canal-like reaches. Ponds, mud.	Larvae, mussels.	Lobworm, bread, maggots.	10–18 in. 2–7 lb	Fair
Trout	Yellowish-colour, black and red spotted back and sides, forked tail.	Fast running streams, gravel beds.	Worms, slugs, crustacea, flies.	Worm, fly.	10–12 in. 1–2 lb	Highly esteemed

tied in muslin or put in a punched tin. Other traps are made with straight sides, and there are models for catching different fish, including eels and pike. Fish traps are illegal except when allowed by ancient right and accepted by the River Board.

NIGHT LINES. Baited lines are left out overnight attached to pegs in the bank, and the catch drawn in in the morning. Or lines are tied to overhanging branches so that the wind moves the trailing fly or bait in an attractive way on the stream. Or a line is tied to two stones, with several baited hooks attached on auxiliary lines; dropped into a stream, its position is marked and the line retrieved later with a hooked stick. The use of such lines depends upon River Board regulations, and is generally illegal.

HUNTING WITH GAFF OR SPEAR. Eels are hunted with a special barbed, three-pronged eel-spear called a pritch. Other large fish are sometimes speared or gaffed after being atttracted to the surface by a light. (It is illegal to direct a light on to the water to take fish.)

TICKLING TROUT AND TENCH. Having made out the position of a fish and lying down on the bank above it, the hunter slowly slips his bare arm into the water with his hand as close as possible to the fish. Passing his finger under the belly, he gently rubs it, when the fish will slowly rise, enabling the hand to position itself for the decisive grab: finger and thumb inserted into the gills and the fish suddenly jerked up and out.

SNARING PIKE WITH WIRE LOOP. Richard Jefferies (op. cit.) describes how jacks were taken with a long slender ash stick to which was attached a loop and running noose of thin copper wire. Acting as quietly and invisibly as possible, the hunter, having marked his prey in the shallow, would extend the rod across the water 3 or 4 yards upstream, letting it swim down with the current until it reached the pike, when it was slipped over its head, past jaws, gills and first fins until it came to a place about one-third of the way down the fish's

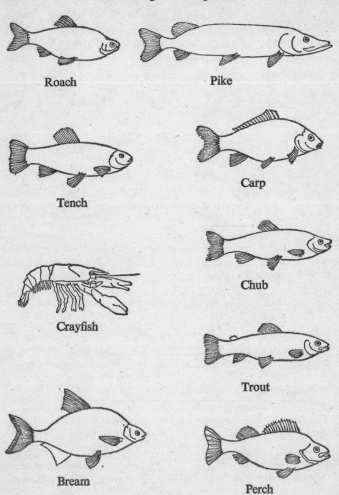

Roach

Pike

Tench

Carp

Crayfish

Chub

Trout

Bream

Perch

Fig. 15

length, when the hunter cleared his arm and, with a sudden jerk, lifted the fish right out of the stream on to the sward. The effect of this was to momentarily paralyse the fish, sometimes to kill it. Lacking a wire, says Jefferies, the skilful hunter would make do with a peeled and twisted withy, a noose and running knot being made in the stick itself: but this required rather greater dexterity than the wire. Besides pike, tench and large roach were sometimes snared in this way. It is now illegal.

NETTING FISH. Jefferies also describes how a net would be stretched from bank to bank and watched by one man, while his fellow walked up the brook 30 or 40 yards and drove the fish down the current towards it by thrashing the water with a pole. When the cork floats bobbed under water the watcher knew that a shoal of fish had run into it, drew the string and made his haul.

OTHER METHODS. A sledgehammer brought down hard on a rock in a shallow stream will set up shockwaves in the water, stunning any nearby fish, which can then be netted or picked up by hand. A temporary dam, built above a pool, will sometimes sufficiently empty the pool of water for any trout it contains to be caught by hand.

A way of taking trout from a small brook so overgrown with briars and thorn that it is impossible to use a rod and line, is to lower a worm on some heavily weighted extrastrong tackle threaded through an old brass curtain-rod. The bait is pulled up to the end of the tube, the tube pushed through the bushes, and the bait taken to the water by the weight. When hooked, the fish is pulled in until it is struggling at the end of the rod. The fisher then simply walks backward and hauls in the catch.

The refraction of light in water, which makes it difficult to gauge the true position of a fish, can be obviated by the wearing of polaroid glasses. Through these, fish can be seen which are invisible to the naked eye.

Cooking

BARBEL. Keep alive in fresh water tank 3–4 days before cooking. Score, and soak in oil for 30 minutes. Season, and broil each side 8–10 minutes over moderate heat. Serve with butter blended with lemon juice, chopped parsley, pepper and salt.

CARP. Soak well in running water, then in weak vinegar and salted water. May be baked, roasted, broiled, fried or stewed, or cooked in wine.

CHUB. As for carp.

EEL. Eels are killed with a knife thrust through the base of the skull, or by knocking their heads hard on a stone. Skin an eel by holding its head, covered with a cloth, in one hand and cutting round the neck with the other. Turn the skin down an inch, grip it and strip it off. Chop off head and tail of the skinned eel, open the throat with a knife-point and make a cut at the navel, then push the gut with a skewer from the small hole in the navel through the opening at the throat; wash and dry before cooking. Eels may be made into pies, into soup, or boiled, fried or stewed. To make eel pie, cut the skinned eel into pieces, season with pepper and salt and place in a greased pie-dish. Cover with milk and water, and a flaky pastry crust, and bake until brown.

GRAYLING. Treat as trout.

GUDGEON. Remove gills but leave scales. Cook in deep fat and serve with lemon.

PERCH. To remove the scales, dip the fish for 3 minutes into boiling water, then scrape. Wash the fish in warm water, clean it, cut out fins and gills. Cover with boiling salted water for 10–20 minutes and serve with sauce. Or may be cooked like trout.

PIKE. Pike are best baked. In *The Compleat Angler*, Izaac Walton gives instructions for roasting a pike on a spit thrust through the mouth and out at the tail, its flesh being prevented from falling off while cooking by half a dozen laths or split sticks tied lengthways along the body: '. . . let him be roasted very leisurely, and often basted with claret wine and anchovies and butter mixed together, and also with what moisture falls from him into the pan.' If to be baked, the pike is cleaned, wiped with a damp cloth, placed in a baking-dish and covered with a mixture of sour cream, grated cheese, butter, salt and pepper, baked for 30 minutes and served garnished with watercress.

ROACH AND RUDD. Can be stewed, boiled or broiled, but are not recommended.

RUFFE. Treat as perch.

TENCH. The best-tasting tench come from clear fresh waters; otherwise the taste may be muddy. If in doubt, keep the live tench in a fresh-water tank for 3 or 4 days before eating. Tench should be washed and soaked in salt water, the gills first being removed. Boil, steam, bake, fry or grill and serve with a butter sauce. To bake: sprinkle the fish with lemon juice and leave for 1 hour. Melt 3 oz butter in a baking-dish and put the fish in it. Baste it, sprinkle with salt and pepper and add 2 chopped shallots. Cover with greased paper and bake gently for 25–35 minutes.

TROUT. Clean and cook soon after catching. To bake: put the whole cleaned fish in a shallow tin or dish, cover with greased paper, bake in a hot oven and serve at once. To broil: split the trout down the back, season with salt, pepper and lemon juice and rub with oil. Cook under a red-hot grill or over charcoal for 5–10 minutes. To fry: split fish and remove the bones; dip in beaten egg and breadcrumbs and fry in hot fat. Garnish with fried parsley.

MINNOWS. These despised small fishes may be easily caught in quantity and cooked like whitebait – that is, they are tossed whole into a floured cloth and then emptied into the frying-basket of a deep pan, where they are cooked rapidly until crisp. Serve with lemon juice. Izaac Walton says of them: '... in the spring they make of them excellent minnow-tansies; for being washed well in salt, and their heads and tails cut off, and their guts taken out, and not washed after, they prove excellent for that use; that is, being fried with yolks of eggs, the flowers of cowslips, and of primroses, and a little tansy: thus used, they make a dainty dish of meat.'

The only large fresh water crustacean is that miniature lobster, the crayfish (Fr. *écrevisse*), a gastronomical delicacy recognized in France if not here, and at its tastiest between May and August. Crayfish make burrows by the sides of fast-flowing streams in chalk and limestone districts (they need lime for their shells) and hide under large stones in the river bed, emerging at night to feed. They can be caught by wading upstream with a net, overturning any large stones, and when one is found, prodding it with a stick when it will dart backward – into the net; or by finding their holes and pulling them out. Or they can be trapped in wire baskets baited with kipper or fish or meat offal. Prepare by washing well, removing intestinal tract from under the tail with a knife-point, and then boiling for 10 minutes in salted water. Take the boiled crayfish, cut in half lengthwise, remove the meat and cut into cubes; from this point, several recipes come into play.

FOR CRAYFISH CARDINAL, made with two crayfish, melt 4 oz butter in a saucepan over a low heat, add crayfish, 2 teaspoons chopped chives and 2 tablespoons cognac. Gently mix, and then pour hot Hollandaise sauce over it, continuing to stir. Serve hot in the preheated shells.

CHAPTER TWENTY-TWO

EDIBLE WILD PLANTS

Throughout summer and autumn the countryside of Britain produces a variety of edible vegetation. The wild vegetable larder is sparse in concentrated starches and proteins, though fairly plentiful in greens and salads. Nuts and fruits are to be found in the autumn; there are a few edible roots. In winter there is almost nothing. Yet nature does provide a number of edible substances which only require finding and gathering to give materials for salads, soups and stews, flavourings, jams and wines, 'teas' and 'coffees', as well as fodder for beasts.

NUTS. The nuts of the hazel are found on roadside bushes and low trees in spinneys and copses in late September. They should be spread out on paper or sacking in the sun for a day or two, when the dry husks can be removed and the nuts packed in sealed containers. Sweet chestnuts are collected from the ground where they fall in October. Stripped from husks, they are spread out to dry and packed in sand or sawdust. They may be boiled, roasted, or ground into a flour for making a kind of bread. Beechnuts or mast are sweet in taste but contain traces of a poisonous narcotic called fagin, which disappears after drying: may be ground into flour.

Acorns are collected in October, skinned, chopped small, oven-dried and stored in tight containers, when they may be ground into a flour as required. Alternatively, crack their shells, soak for 24 hours, changing the water frequently. Remove shells and boil the kernels, changing water every 2 hours until the bitter taste has gone. Dry and use as flour.

OIL FROM NUTS. A very sweet olive-like oil may be expressed from beechnuts, and a bland, fixed oil from hazel nuts, as well as from the cultivated walnut.

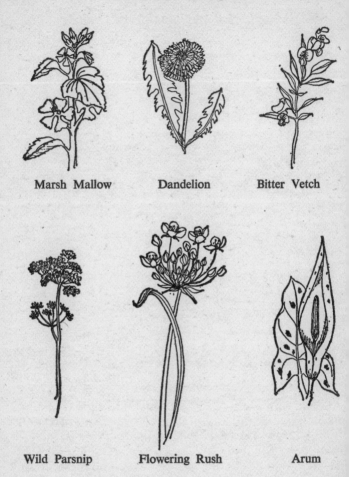

Marsh Mallow   Dandelion   Bitter Vetch

Wild Parsnip   Flowering Rush   Arum

Fig. 16

NUTS AS FODDER. After the extraction of oil the residue of the nuts may be mixed with other foods and fed to cattle. Acorns make fodder for pigs, as does beechmast; the whole acorn may be coarsely ground, the meal then being boiled or scalded and mixed with cattle beet or other fodder. The horse-chestnut, unfit for humans, may be fed to cattle, sheep and poultry after first removing the bitter taste by soaking the partly crushed nuts in cold water for 8 hours, boiling for 30 minutes, grinding or crushing to a meal and mixing with other meals in the proportion of about 40 to 60.

ROOTS. From May to mid-July the Earth-nut (Pig-nut, Earth-chestnut), a near relative of the Cow Parsley and Hedge Parsley with a similar flat-topped spreading flower, may be dug up when found in any quantity in dry woods and grassland. The roots are wholesome and nutritious. Children eat them raw, but they are delicious roasted or boiled, and may be used in soups. Other edible roots which may be boiled and mashed are those of the Wild Parsnip, the Dandelion, the White Water-lily, the Marsh Mallow, the Bitter Vetch, the Silverweed, the Flowering Rush and the common Arum, better known as Cuckoo-pint or Lords and Ladies. The poison in arum root is removed by boiling, when a nourishing starchy substance results, called Portland Sago.

FRUITS AND BERRIES. Crab apples – inedible except as jelly. Sloes are the fruit of the blackthorn; they are inedible but make good wine. Bullaces grow on small trees larger and less thorny than the blackthorn; unlike sloes they can be eaten in pies and tarts: pick after first frost. Blackberries, in conjunction with apples, make pies, tarts and jam. Elderberries are barely edible on their own but make good wine. The Rose-hip is made into a vitamin C-rich syrup or a wine. The Rowan or Mountain-ash has berries which may be made into jellies or wine. The Sea-buckthorn or Sallow Thorn has orange-coloured berries which make a slightly acid jelly. The Blaeberry or Bilberry, which is gathered in August

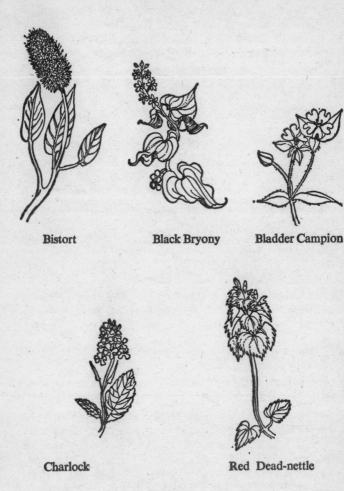

Bistort   Black Bryony   Bladder Campion

Charlock   Red Dead-nettle

Fig. 17

from heathland, is used in tarts or to make jam. The Cranberry is used for tarts and sauces. The Cloudberry may be used in several ways. Also edible are: the Dewberry, and the berries of the Whitebeam, the Guelder rose and the hawthorn which in spite of their insipid taste can be made into jellies and wine with plenty of sugar. The wild varieties of Cherry, Blackcurrant, Raspberry, Strawberry, Red Currant and Medlar are also worth gathering.

LEAVES AND STEMS. The following are edible in fresh salads: Beech leaves, Brooklime Speedwell, Catsear, Chicory, Chickweed, Corn Salad, Corn Sow-thistle, Dandelion leaves, Garlic Mustard (Sauce Alone, Jack-by-the-Hedge), Goatsbeard, Greater Prickly Lettuce, Hairy Bittercress, Hawthorn leaves, Lady's Smock, Land Cress, Lime leaves, Rose-bay Willow-herb leaves (treat young shoots as asparagus), Rough Hawkbit, Salad Burnet, Sea Purslane, Wall Lettuce, Watercress, Wood Sorrel (blossoms), Yarrow. Avoid watercress gathered just downstream of a meadow grazed by sheep as there will be danger of liver-fluke.

These may be cooked as greens: Bistort (Snakeweed), Black Bryony, Bladder Campion, Charlock, Chickweed, Common Comfrey, Common Hop, Common Mallow, Common Sorrel, Fat Hen, Good King Henry, Goosegrass, Ground Elder, Halberd-leaved Orache, Henbit, Hogweed (Cow Parsnip), Red Dead-nettle, Shepherd's Purse, Sow Thistle, Stinging Nettle, White Dead-nettle, Yellow Archangel. In general these leaves are cooked like spinach – they are picked when young and tender (with nettles only the tips are used), washed and cooked gently in a saucepan without more water but with a wedge of butter and some seasoning. They may also be made into soups. In addition, the curled fronds of Bracken may be treated like asparagus.

These plants have edible stems: Burdock; Marsh Samphire; Rock Samphire; Sea Kale; Sow Thistle.

TEA AND COFFEE SUBSTITUTES. The best substitute for tea in a time of shortage is a caddy full of the dried and crushed young leaves of Blackthorn, which were once used commercially in blending tea, until the practice was stopped by law. Lime flowers may be dried and used, either by themselves or mixed with Indian or China tea. They relieve headaches and indigestion. Dried Elderflowers make a delicate aromatic tea and enhance the flavour of Indian tea if added in the proportion of about 1 to 20. The green leaves of Peppermint make a pleasant tea, and are a remedy against flatulence. An ounce of Hops added to half a pound of tea makes it more digestible and gives it tonic properties and a slightly bitter flavour. The dried flower heads of Heather make a good tea. Also usable are: the leaves, flowers and stem of Agrimony; the flowering tops of Mugwort; the tips of Lemon Balm; Betony, dried and powdered; the dried flowering tops of Catmint; Lovage; Meadowsweet; Spearmint; Sage, and Centaury.

There are three principal stand-ins for coffee: acorns, and the roots of Chicory and Dandelion. The acorns must be ripe when gathered. Skin, cut into four and roast in a hot oven until they become cinnamon-brown in colour; grind in a mill and put away in a tight-closed tin. The roots are washed, roasted in a slow oven until brittle, and ground in a coffee-mill. The ripe seeds of the Wild Iris or Yellow Flag, roasted and ground, also make a fair coffee, as do the seeds of Goosegrass or Cleavers, a variety of Bedstraw.

SUGAR FROM TREE SAPS. Three British trees may be tapped like the Canadian Maple for their sugar-bearing sap: the Greater Maple or Sycamore (*Acer pseudo-platanus*), the Birch, and the cultivated Walnut. The sap of each may be converted into wine by fermentation, or it may be evaporated to produce a syrup or a solid sugar. But it takes 40 gallons of sap to make 1 gallon of syrup, while the quantity of sap yielded by one tree in a season will seldom exceed 24 gallons, and this amount of extraction may endanger the tree. The best trees are wide in the bole and with a good crown growth, a large full crown of leaves indicating large amounts of sweet sap.

Tap the tree at the time of its fullest sap-flow, which is

usually reckoned to be in early March. Sap flow usually follows an alternation of warm days and frosty nights. To test for sap, break off a twig, when sap, if present, will exude. Tie a string to the twig guiding the drops of sap down into a bucket or a sealed plastic container. A gallon of sap will take 4–5 days to collect. Alternatively, make a $\frac{1}{2}$-in. deep incision in the tree, slanting upwards, and gradually increase this to 2 in. Insert a spout of elder through which the sap can flow into a bucket or a plastic bag hung beneath it in such a way that dirt and insects are kept out. Strain the liquid and boil it soon after collection to forestall fermentation. Use a shallow vessel, maintaining the depth of syrup at draw-off point at $\frac{1}{2}$ to 1 in., and draw off the syrup as soon as it reaches its boiling temperature of 7°F above the boiling point of water. The syrup must be bottled hot, at 180°F or more.

Syrup is crystallized by raising its boiling point to 7·5°F or more above the boiling point of water; this releases more water and supersaturates the syrup which, when it cools to a room temperature of 68°F, will precipitate the excess sugar in the form of crystals. To help crystals to form, stir the hot solution continuously while it is cooling: this mixes the crystals throughout the thickened syrup and they grow in size and number.

Sugar may also be extracted, by boiling, from the Sugar Beet and Mangold.

EDIBLE FUNGI. Fungi are a class of vegetables which, lacking chlorophyll, have to obtain their organic food ready made, either by using dead plant and animal products or by attaching themselves to living animals and plants as parasites. Mushroom is the name given to the fruiting body of the larger, fleshy fungi, particularly (though not exclusively) to those bearing gills, and more especially to those which are sought as food. Mushrooms vary in edibility from the delicious through the palatable and the unpleasant to the nauseous and deadly poisonous, and both esculent and poisonous varieties are to be found within the same genus. (The use of the word toadstool is meaningless, and should be abandoned.) The food value of the mushroom is equal to that of most vegetables except that it contains rather more protein and makes up for the absence of vitamin C by possessing a large amount of vitamin B and the antirachitic vitamin D.

Some fungi are to be found from April onwards, but September and October are the main mushroom months. Mushrooms not wanted for use now may be strung on thread, dried slowly, and stored in glass jars, or made into ketchup.

COLLECTING MUSHROOMS. Since it is impossible to get to know all the thousands of species of mushroom, the best course is to learn to recognize on sight about a dozen of the most common and flavoursome esculents, together with the commonest and deadliest of the poisonous kinds, of which there are few. There is no rule of thumb for sorting out poisonous from harmless varieties, and knowledge must be got by collecting specimens, taking them home and comparing them with the descriptions and pictures in a handbook such as *Wild Mushrooms* by Linus Zeitmayr (Muller: London) which covers both edible and poisonous fungi.

POISONOUS FUNGI. Because the four most poisonous fungi all belong to the same genus, the *Amanita*, they are not hard to recognize. They are: The Death Cap (*Amanita phalloides*), The Fool's Mushroom (*A. verna*), Destroying Angel (*A. virosa*), and False Death Cap (*A. citrina*), this last now known to be harmless, but best avoided because it looks so like the real Death Cap. It is their likeness to the common field mushroom which makes these fungi dangerous to the picker. Two other *Amanitas* are less dangerous because more easily recognizable: the Fly Agaric (*A. muscaria*) and the Panther Cap (*A. pantherina*), both of which are rather decorative, with raised white spots on their red or brown caps.

AMANITAS are easily distinguished from edible mushrooms by these signs:

Amanitas	*Mushrooms*
Always have white gills.	May have white, pink, lilac, brown gills.
In early stage of growth have an egg-like look, by reason of their being enveloped in a 'universal veil'.	Have no universal veil, only a partial veil leaving a ring below the cap when it is torn.
With growth, this veil tears, leaving the base of the stem sheathed with membrane called the 'volva'. All *Amanitas* have this volva.	No mushroom has a volva.

One other fungus may cause trouble, the Yellow-staining Mushroom (*Agaricus xanthoderma*), a true Mushroom which with pink and brown gills and lacking a volva looks like the edible Horse Mushroom; it gives itself away by becoming bright yellow where rubbed or cut, especially at the base of the stem.

The rules for collecting fungi, therefore, resolve themselves into two: (1) reject any Mushroom which has white gills and a volva at the base of the stem; and (2) reject any fungus resembling a Horse Mushroom if, on cutting the base of the stem, it turns bright yellow. Most esculent fungi can be recognized easily with practice, and mistakes will at the most have uncomfortable rather than dangerous results.

Descriptions of edible fungi

COMMON FIELD MUSHROOM (*Agaricus campestris*). The best known of the wild species, having a whitish silky cap up to 4 in. across and gills that are first white, then pink, purple and blackish. White, smooth stem. Partial veil tears away to form annulus, or ring. Resembles the cultivated mushroom of the greengrocer's shop in appearance. Spores nearly chocolate brown.

HORSE MUSHROOM (*A. arvensis*). Similar in appearance to the above, but larger in size, and gills not pink.

CEP (*Boletus edulis*). Found on the ground in woods in autumn. Has a large, brown, rounded bun-like cap 4 to 6 in. across and a short, thick, often slightly bulbous stem $2\frac{1}{2}$ to

6 in. long, often yellow-brown and marked with a network of raised lines on upper part. Its spores (which are yellow-olive) are produced in tubes rather than on the sides of gills. The flesh is thick and white. To dry for storage and use out of season, wipe ceps with a clean dry cloth, thread a needle with strong linen thread and string them on it. Hang in a cool oven (240°F) and after 24 hours remove and pack into glass jars. Cooking: Remove tubes and cut up flesh into slices. Fry in olive oil and season to taste.

BLEWITS (*Tricholoma personatum*). Grows among grass in pastures, often in large rings. Has a thick, rounded cap of pale brown colour with a smooth surface, from 2½ to 5 in. across. A short, thick, fibrous stem, sometimes bulging at the base. Off-white or faintly pink gills often detached from the stem. The flesh is thick and firm, becoming spongy. Pink or lilac spores.

WOOD BLEWITS (*T. nudum*). Grows in woods, singly or in groups, in October/November. Cap is 2½ to 7 in. across, at first convex, becoming flat, lilac or lavender in colour, including gills, which are rounded behind, sometimes nearly free of stem. Stem is 1 to 3½ in. long and up to 1¼ in. thick, solid, bulging at base, at first blue, fading to lilac. The spores are flesh coloured.

ST GEORGE'S MUSHROOM (*T. gambosum*). So called because it appears about 23 April, in pastures, often in rings. It is of a sturdy appearance, buff in colour, with a smooth cap 3 to 4 in. across, rounded and fleshy, with an incurved, wavy margin. Off-white to buff stem, short and stout, with white, downy apex. Gills are white to pale buff, with white spores. Cook slowly with a little butter, serve with white sauce.

CHANTARELLE (*Cantharellus cibarious*). Look for it in summer and autumn on the ground in woodland where it occurs singly or in groups and small clusters. It is crome- to egg-yellow in colour with a smooth cap at first convex but later flattened and hollowed and at last funnel-shaped. Thick, firm flesh, with thick blunt-edged gills, sometimes fold-like, extending down the stem, of much the same colour as the cap. When fresh, smells of apricot. Spore is pinkish buff. Cooking: Wash very thoroughly, slice and cook in butter in a covered frying pan. When tender add parsley butter and a tablespoon or two of clear stock to form a little sauce. Mature specimens may be tough and will require slow cooking or stewing.

SAFFRON MILK CAP (*Lactarius deliciosus*). Found in damp woods and boggy places in September/October. The cap, up to 5 in. across, is orange or red, with concentric zones of deeper colour, later paling to grey or grey-green, at first convex then becoming funnelled. The orange stem becomes green-spotted, is short and pithy, becoming hollow. When the gills or flesh of the cap or stem are broken a bright orange juice exudes, turning green on exposure. The flesh is pinkish orange; the spores pinkish-buff. Cooking: Grill rapidly and serve with a little butter, or use for flavouring soup. Does not dry well.

PARASOL MUSHROOM (*Lepiota procera*). Grows singly or in groups in open woods or pastures during August/September. At first resembling a drumstick, this large fungus expands to a parasol-shape up to 8 in. across, with a marked central boss. The stem is very tall, tapering from its bulbous base, the cap covered with widely spaced brown scales, with white flesh beneath. The annulus can be moved up and down the stem; it is large and thick with brown scales underneath. The gills are white and free from the stem; the spores are white.

SHAGGY PARASOL (*L. rhacodes*). More stubby than the Parasol, this one has a more rounded cap, with a less pronounced central boss. It grows on heaps of garden compost, and in woods. The cap, 4 or 5 in. across, is pale, the scales large and shaggy, with a tendency to turn up. The smooth white stem swells suddenly into a bulb at the base and tapers upwards to a height of 6 to 8 in., and it has a movable fringed ring. The gills are white and free from the stem; the flesh of cap and stem turns reddish when cut.

MOREL (*Morchella esculenta*). Found in spring, in forest-clearings and in hedges, orchards and pastures. Its distinctive sponge-like head is greyish brown to yellowish-brown, more or less elongated, its surface covered with irregular, rather elongated pits, and its stalk being fragile, whitish in colour and hollow. Cooking: Wash before use to remove all grit. Stuff with minced meat and bake in a casserole, or use as flavouring for soups and stews. Will dry.

GIANT PUFF-BALL (*Calvatia gigantea*). May be found in open woods, pastures and fields in autumn – a more or less globose growth, with a soft white leathery surface, attached to the ground by a short cord of root. Cooking: Cut in slices and fry.

When all these varieties are known at sight, and have been often gathered and eaten, it is safe to move on to other edible fungi, although those named above are by common consent the most worth gathering.

Cooking: Mushrooms generally can be baked in the oven, fried very slowly (using not too much butter), and grilled whole. They can be used in combination with other vegetables to make a dish (mushrooms and chestnuts, and mushrooms, tomatoes and green peas are good combinations), and as a flavouring for soups and stews.

TRUFFLES. The truffle is a black, warty, edible fungus which lives parasitically on the ailing roots of trees, about a foot under the ground. It prefers a light chalk soil, and the roots of the oak, but has also been found in association with beech, birch, chestnut and hornbeam roots. It is nearly globular in shape, and somewhere between a walnut and a large potato in size. It cannot be cultivated or even encouraged, and in its natural state is elusive. Its presence beneath a tree may be shown by the dying away of the herbage on the ground in a circular pattern around the tree which spreads gradually outwards; but even with this sign the precise position of the truffle must be found so that it can be dug up neatly without harm to a possible future crop. Sometimes the truffle betrays its presence by a faint cracking of the earth crust immediately above it, or on still winter days by the columns of gnats which hover just over the ground where it lies, attracted by its fragrance. The truffle is useless

for food until it ripens, and so should only be dug for in the
months between November and March.

COOKING. Wash and brush in several waters to remove all
grit. Wrap separately in buttered paper and bake in a hot
oven for one hour. Truffles are most commonly used in
pâtés.

MUSHROOM POISONING. Avoid giving alcohol to the
patient. In all cases other than *Amanita* poisoning use an
emetic and purgative to empty stomach and bowels of their
contents. The antidote is oak bark in milk, which must be
given before the poison has passed into the bloodstream.
Prepare a decoction by boiling one handful of fresh or dried
bark in 2 pints of milk. Give the patient plenty of sweet hot
tea to drink, cold tea in the case of acrid Russula or Milk
Caps; apply hot fomentations to abdominal pains.

CHAPTER TWENTY-THREE

BRIEF HERBAL

That the herbs of the field have bio-chemical properties which are capable of acting on the human organism in both beneficial and harmful ways has long been known.

Herbal medicines fall traditionally into ten categories corresponding to ten main types of physical need or discomfort.

ALTERATIVES. Medicines which alter the condition of the body and improve general health. These include such blood-purifying herbs as Burdock, Red Clover, Stinging Nettle, Yellow Dock.

ASTRINGENTS. Medicines which produce contraction of the muscular fibre, and in particular counteract looseness of the bowels. Blackberry, Nettle, Oak-bark, Plantain, Witch Hazel.

CARMINATIVES. Medicines which aid digestion and relieve flatulence and stomach-ache. Angelica, Balm, Barberry, Burdock, Clover, Wormwood.

DEMULCENTS. Medicines which by a bland and soothing action allay irritation, especially in the case of inflammations of stomach and lungs. Agrimony, Comfrey, Linseed, Marsh Mallow.

DIAPHORETICS. Medicines that are able to provoke perspiration, thus sweating out colds and fevers. Balm, Catmint, Yarrow.

DIURETICS. Medicines which promote the discharge of urine. Asparagus, Earth-nut, Dandelion, Pellitory of the Wall, Wild Carrot, Bladderwrack.

EMETICS. Doses that induce vomiting. Mustard, salt.

EXPECTORANTS. Medicines that clear the chest, throat and lungs of phelgm. Horehound, Lungwort.

LAXATIVES: Preparations that loosen the bowels. Chickweed, Dandelion, Linseed, Rhubarb.

TONICS. Medicines which give tone and vigour to the system. Balm, Blackberry, Camomile, Centaury, Gentian, Hops, Sage.

Herbal medicines, called simples when used singly, are normally taken internally in a decoction of the fresh or dried leaves, stems, flowers, berries or roots of the plant in question, often of the whole plant. Decoctions are made by covering the plant with boiling water and leaving it near the fire overnight to steep, the resulting liquid being poured off and used or bottled, the solids being thrown away unless required for use as a poultice. The following list is not meant to be more than a rough guide to ailments and some of their accepted remedies.

Anaemia Dandelion, Nettle

Asthma Coltsfoot, White Horehound, Great Mullein, Willowherb

Biliousness Dandelion, Groundsel, Strawberry

Bladder trouble Couch-grass, Dandelion, Horsetail

Boils Chickweed, Ground Ivy, Yarrow

Bronchial trouble Clover, Coltsfoot, Ground Ivy, Lungwort, Sunflower seeds, Linseed

Catarrh Clover, Ground Ivy, Elder

Chapped skin Groundsel, Silverweed (ext.)*

Chilblains Onion and salt (ext.)

Colds, chills Dead Nettle, Yarrow, Elder

Constipation Chickweed, Dandelion, Groundsel

Coughs Clover, Ground Ivy, Linseed, Speedwell, White Horehound

Cramp All-Heal, Calamint, Cowslips, Mustard, Pennyroyal, Saxifrage (ext.)

Cystitis Couch-grass, Marsh Mallow

Diarrhoea Blackberry leaves, Oak bark, Comfrey root

Dyspepsia Dandelion, Ground Ivy, Mint, Horseradish

Eczema Dandelion, Walnut leaves and bark, Watercress (ext.)

Eye lotions Chickweed, Ground Ivy, Eyebright, Fennel, Meadowsweet

Flatulence Clover, Dandelion, Ground Ivy, Wormwood, Calamint

Foot discomfort Silverweed, Bedstraw (ext.)

Gallstones Couch-grass, Silverweed, Sow Thistle

Gout Cinquefoil, Couch-grass, Dandelion, Ground Elder, Ground Ivy, Nettle, Carrot, Chicory (root)

Gravel Couch-grass, Dandelion, Nettle, Dropwort, Furze

Insomnia Lettuce, Privet

Jaundice Couch-grass, Ground Ivy, Yarrow, Thyme, Lady's Mantle

Kidney trouble Clover, Couch-grass, Dandelion, Thyme, Parsley, Sage

Liver trouble Dandelion, Dock, Maple leaf or bark, Sage

Neuralgia Poppy heads, Camomile

Obesity Bladderwrack, Chickweed
Piles Celandine, Elder, Blackberry, Witch Hazel
Pulmonary complaints Ground Ivy, Hedge Mustard, Borage
Rheumatism Couch-grass, Ground Elder, Nettle, Bladderwrack, Celery Seed, Comfrey, Chicory root
Sciatica Ground Elder, Ground Ivy
Skin eruptions Dandelion, Horsetail (ext.)
Sprains and bruises Witch Hazel (ext.)

Stomach upset Chickweed, Oak bark, Lemon Verbena
Stone in bladder Lady's Mantle
Throat, relaxed, sore, ulcerated Herbs of *potentilla* family, Nettle, Blackcurrant
Ulcers and wounds Comfrey, Marigold, Self-heal, Southernwood, Wintergreen, Herb Robert, Loosestrife (ext.)
Varicose veins Daisy (ext.)
Warts Milk of Dandelion, Celandine, Lining of Broad Bean shell (ext.)
Whooping cough Clover

* (ext.) indicates external application.

Some further uses of herbs

ELDER. A bush of common elder outside a kitchen window keeps away flies.

SAGE. Some leaves of sage burned on a shovel or plate destroy odours.

MARIGOLD. The petals of the ray florets are used to flavour buns, soups, stews. May be used to colour butter and cheese.

PARSLEY. A purifier of the breath after smoking or eating onion or garlic.

MARJORAM. Lay a sprig of marjoram across a milk jug to keep milk fresh.

SLOE. The juice of a ripe sloe is an effective marking-ink: stick nib into the fruit and write.

LADY'S BEDSTRAW. Sometimes called Cheese Rennet because it has the property of curdling milk.

SOAPWORT. Both root and leaves contain saponin. 'Bruised and agitated with water, it raises a lather like a soap, which easily washes dirty spots out of clothes.' – Culpeper. Some species of the genus Gypsophila contain much saponin, which is also to be found in horse-chestnuts.

TANSY, RUE AND MINT. All these deter flies.

POISONOUS HERBS: Anemone; Aconite; Belladonna (Deadly Nightshade); Autumn Crocus; Foxglove; Hemlock; Henbane; Thornapple.

WEIGHTS, MEASURES AND TEMPERATURES

BRITISH MEASURES AND METRIC EQUIVALENTS

Length
1 inch = 2·5400 centimetres (cm)
1 foot = 0·3048 metre (m)
1 yard = 0·9144 m
1 rod = 5·0292 m
1 chain = 20·117 m
1 furlong = 201·17 m
1 mile = 1·6093 kilometres (km)

Surface or Area
1 sq. inch = 6·416 cm²
1 sq. foot = 0·0929 m²
1 sq. yard = 0·8361 m²
1 acre = 4,046·86 m²
1 sq. mile = 259·0 hectares

Volume
1 cu. inch = 16·387 cm³
1 cu. foot = 0·0283 m³
1 cu. yard = 0·7646 m³

Capacity
1 pint = 0·5683 litres
1 quart = 1·1365 litres
1 gallon = 4·5461 litres
1 bushel = 36·369 litres

Weight
Avoirdupois
1 ounce = 28·350 gramme (g)
1 pound = 0·4536 kilograms (kg)
1 stone = 6·3503 kg
1 cwt = 50·802 kg
1 ton = 1·0161 tonnes

TEMPERATURE

Boiling point is 100° Centigrade (Celsius) and 212° Fahrenheit. Freezing point 0°C and 32°F. To reduce Fahrenheit to Centigrade subtract 32 degrees and multiply by 5/9.

°Centigrade	°Fahrenheit	°Centigrade	°Fahrenheit
100	212	36·7	98
95	203	35	95
90	194	32·2	90
85	185	30	86
78·9	174	26·7	80
75	167	25	77
70	158	20	68
65	149	15·5	60
60	140	12·8	55
55	131	10	50
52·8	127	7·2	54
50	122	5	41
45	113	1·7	35
42·2	108	0	32
40	104		

A COUNTRY GLOSSARY

ACRE area of land containing 4,840 sq. yards (imperial acre), divided into 4 roods each of 1,210 sq. yards (see *chain*)

ADZE axe-edged, hoe-shaped tool for rough-hollowing timber

AGGREGATE any coarser material mixed with cement to form concrete

AGISTMENT arrangement involving the letting of grazing rights

ANAEROBIC change effected by the agency of bacteria living in the absence of free oxygen

ANNUAL plant that completes its life-history in one year

ARABLE land under cultivation and used mainly for cropping

ARTESIAN WELL well sunk in a valley or 'basin' under conditions which ensure that water rises and discharges at the surface with some force

BALE straw or hay compressed and bound in rectangular block

BARM froth of fermenting liquor

BARROW male pig castrated after period of use as a boar

BASTE to pour fat over (a joint of meat)

BAT stout stick

BATTER inward slope from perpendicular

BAVIN a *faggot*, q.v.

BEE BREAD stored pollen in comb

BEETLE, BITEL large wood mallet used for driving stakes and splitting timber

BEEVES cattle, oxen

BESOM a rough broom of twigs

BIENNIAL plant which completes its life history in 2 years

BIKE wasp nest

BILLET thick piece of split log

BIN container for grain. To find capacity of a bin in bushels, multiply length by width by depth in feet, and multiply of 0·78 (roughly, 0·8)

BIO-GAS marsh gas, methane

BOLE trunk of tree between soil level and lower branches

BRAN outer skin of wheat grain

BRASSICA genus of Cruciferous plants, including cabbage, turnip, rape, etc.

BREWER'S GRAINS spent grains after extraction of malt, fed to stock

BRINE salt in water solution. 1 per cent salt in brine = 1 oz salt per gallon water; 10 per cent = $14\frac{1}{4}$ oz; 15 per cent = $22\frac{2}{3}$ oz

BROCK swill

BROIL to grill, or cook over hot coals

BROTCHES pegs used in thatching

BULL uncastrated male of the cattle family

BULLOCK castrated male cattle

BUSHEL measure of capacity used especially for grain, measuring 2218·2 cu. in., and containing 80 lb or 8 gallons of water. In dry measure the bushel contains 8 gallons or 4 pecks, and 8 bushels go to the quarter. A bushel of English wheat, rye and maize contains 60 lb, barley 50 lb, oats 39 lb

CABER thin tree trunk or rafter

CALLUS scab that forms over a cut surface

CAMPDEN TABLET sterilizing agent (potassium metabisulphite) used in fruit preserving and wine making

CAST (of bow) power of propelling arrow

CATCH-CROP crop grown quickly between two main crops

CESSPOOL system of disposing of sewage by underground storage in tank which must be periodically emptied

CHAFF fine-cut oaten sheaves, the main food for horses

CHAIN 22 yards, or 100 links. 10 chains = 1 furlong, or 'long furrow', the distance an ox could plough without resting, and therefore the most common length of a field. There are 10 sq. chains in 1 acre. If an oblong field is $5\frac{1}{2} \times 6\frac{3}{4}$ chains, to find its area multiply 5·5 by

6·75 = 37·125 chains, or 3·7125 acres = approx. 3¾ acres

CHATS   very small potatoes, fed to pigs

CLAMP   cache of root vegetables stored out of doors

CLEAVE   to split wood along the grain

COMB   a measure equal to 4 bushels

CONCENTRATE   protein and oil-rich rations for stock

CONCRETE   mixture of Portland cement with sand and aggregate of varying proportions:
> For foundations: 1 cement, 8 'all in' aggregate 1½ to ³⁄₁₆ in. For solid floors: 1 cement, 6 'all in' aggregate. For reinforced work: 1 cement, 2 sand, 4 aggregate ¾ to ³⁄₁₆ in.

CONY   rabbit

COPPICE   spinney from which timber is systematically cut for use

CORD   128 cu. ft (of timber); roughly, a pile of logs 8ft × 4 ft × 4 ft

CORDWOOD   tree branches cut from trunk of felled tree

CORM   a solid, swollen underground stem

COULTER   metal cutting edge of ploughshare

COURSE   system of rotation cropping to suit soil of a particular district

CREEP FEEDER   arrangement which allows food to be fed to young pigs to the exclusion of the mother-sow

CROWBAR   iron bar shaped for use as a tool

CULL   to pick out weak specimens of stock

CURTILAGE   yards and premises about the house

DAM   female parent of animal

DEADSTOCK   lifeless objects, as distinct from animals

DEAD WEIGHT   weight of carcass

DECIDUOUS   trees which lose their leaves annually

DECORTICATED   nut meal from which the fibrous cortex has been removed

DOGS   iron stands over which logs are laid for burning

DOLES or NIBS   handles attached to scythe shaft

DOWSE, DOWSER   to locate water underground by means of

involuntary muscular movement communicated to a rod held in dowser's hands

DRAKE   male duck

DRENCH   medicine for animal; to administer such

DREDGE CORN   mixture of cereals and pulses grown for animal feed

DRESS   to clean corn of impurities

DREY   nest of squirrel

DRYSTONE   stone-built without mortar

ENSILAGE   *silage*, q.v.

ESTOVERS   (legal) right to gather wood for fuel

FAGGOT   wired bundle of twigs for burning

FALLOW   land left unsown to recover after cropping

FARROW   litter of pigs

FILLY   young female horse

FINE   to clear a wine of floating particles

FLITCH   side of bacon

FODDER   food supplied to cattle and horses

FROE   woodcutting tool with blade set at right-angles to handle

FURLONG   see *chain*

GAFF   fisherman's hooked pole

GALLON   liquid measure = 4 quarts = 8 pints = 0·1606 cu. ft = 4·546 litres. 1 gallon water weighs 10 lb

GAME   wild animals hunted for sport

GANDER   male goose

GELDING   castrated male horse

GILT   young sow

GOATLING   young female goat

GRIST   malt after grinding

HARDCORE   stone and brick rubble

HARDWOOD   timber from deciduous trees

HAULM   (dried) stem and leaves of long-stemmed plants

HAYSEL   hay harvest

HEAD   (of water) energy owing to height, velocity and pressure

HEAT   period of oestrum in female animal

HEIFER   young cow

HELVE   axe handle

HINNY   hybrid of horse and female ass

HOG   male pig castrated when young

HOGG, HOGGETT   young sheep

HOGGIN   a soft coarse red sand

HOGSHEAD   cask of 52½ imperial gallons

HOVEN   bloat; distention of stomach of grazing animal by gas

HUMUS   decomposed organic matter in topsoil

KERF   cut made by saw-teeth

KIBBLED   coarsely milled (grain)

LACTATION   coming into milk of female animal

LATERAL   side-shoot of plant

LEES   sediment in wine

LEGUMES   plants bearing pods, e.g. beans, peas

LEVERET   young hare

LEY   arable land under grass or pasture

LIGHT   a glazed frame

LITTER   straw for floor

LOAM   the top 6 in. of pasture

LYE   strong alkali solution

MARC   the pressed pulp in cider, wine or oil making

MAST   the fruit of beech, oak, chestnut and other forest trees

MIDDLINGS   coarser part of wheat left after sieving

MONKEY-WINCH   machine with winch and lever used to grub up tree stumps

MORTAR   (soft) lime and sand mixed with water; (cement) Portland cement mixed with sand and water. For bricklaying, use 1 part cement to 4 parts bricklaying sand. Wet bricks before using

MOULD-BOARD   part of plough that turns over the furrow

MUCKING OUT   removing dung from stable, byre, etc.

MULCH   straw, peat or compost laid at the roots of plants to conserve moisture

MULE   neuter hybrid between mare and male ass

MUST   juice expressed from fruit before fermentation into wine

NIGHTSOIL   human excrement in raw state

OSIER   willow which yields rods for basket making

OX   male or female of common domestic cattle, especially a castrated male

PAN   hard layer of soil at level of repeated cultivation

PASTURE   land left under grass for grazing of livestock

PECK   2 gallons or $\frac{1}{4}$ bushel

PECTINOL   pectic enzyme used in wine making to reduce haziness in wine

PELT   skin of small furred animal

PISCARY   (legal) right to fish in common waters

PLUCK   (of pig) the lungs, heart, liver, windpipe and gullet

POKE   a bag

POMACE   crushed apples for cider

PONY   small horse, less than 13 hands high

PORKER   pig of 60–80 lb dead weight

PROVE   allow dough to rise by the action of carbonic acid gas

PUDDLE   mixture of clay and sand; to make water-tight with clay

PULLET   young fowl on point of lay, or just laying

PUNK   touchwood, wood decayed by fungal attack

QUARTER   of wheat, $4\frac{1}{2}$ cwt

QUARTERN   loaf of 4 lb, as if made from $\frac{1}{4}$ stone of flour

RACK   to draw off (wine) from lees

REGISTERED   recorded in herd book

RENDER   to extract by melting

RIPARIAN   of a river bank

ROD 5½ yards (length of ox-goad)

RODENT gnawing animal with front teeth that continue growing throughout animal's life

ROGUEING eliminating inferior specimens, in plant breeding

ROUGH SHOOTING where no keeper is employed and game is not reared of purpose

ROTATION systematic annual change of ground for specific crops

RUBBISH weeds

RUNT the poor pig in a litter

SACK bag; the corn sack contains 4 bushels, but the weight varies according to contents: 1½ cwt oats, 2 cwt barley, 2¼ cwt wheat, 2 cwt 3 stones beans. Allow 4 lb weight of sack

SEPTIC TANK system of disposing of sewage from water-closets through anaerobic bacterial action

SEASON to leave timber in open to dry out before use: allow 1 year for each inch thickness of planks

SCION part of a tree, joined to a root-stock of more vigorous growth in grafting

SCORE 20 lb weight of a pig

SCREEN to put through a sieve

SETT badger's lair

SERVE to impregnate female animal

SHARD gap in a hedge

SHARE blade of plough

SILAGE fodder made by fermentation of green crops in a silo

SIMPLE medicinal herb

SIRE male animal used for breeding

SKILLET saucepan with long handle and three legs for standing in fire

SLAKE to hydrate, as lime

SLURRY solids mixed with water

SNAITH stem of scythe

SOD clump of grass with soil

TO SOIL to feed stock with greenstuff cut straight from field or garden

SOW mature female pig

SPINNEY small area of woodland, kept standing as covert for game

SPIT length of a spade blade

SQUAB young pigeon raised for food

STALLION uncastrated male horse

STEER young bullock

STILTS plough handles

STOOK a shock of sheaves set up in a field

STOOL a stump from which shoots sprout

STORE cattle kept for fattening (or, pigs held in hand for breeding or fattening)

SWATH one sweep with a scythe; the line of mown hay

TALLOW beef or mutton fat rendered in boiling water

TEG sheep in its second year

TETHER to fasten an animal by means of harness, chain and stake

TILT canvas or tarpaulin cover

TINE prong of fork or harrow

TRAIL entrail

TRIVET three-legged stand with a handle, placed in embers to support kettle or pot

TUBER fleshy underground stem

TURBARY (legal) right to dig turf on another's, or common land

VERMIN all obnoxious insects, animals and birds

WEATINGS former term for *middlings*, q.v.

WETHER castrated ram

WITHY coppice willow

WORT unfermented, or fermenting malt

YARD (of gravel, sand) 27 cu. ft

YEALM prepared layer of wet straw used in thatching

INDEX

(wild plants in italic type)